Dad (Watashee)
Thank you for
encouraging me to explore
and become a part of my
heritage. I hope this will help fill
in any gaps for us!   I love you—
Rebecca ☺
Father's Day 2001

A SHEARWATER BOOK

A FACE IN THE ROCK

A FACE IN THE ROCK

The Tale of a Grand Island Chippewa

Loren R. Graham

Illustrations by Abigail Rorer

ISLAND PRESS / Shearwater Books

Washington, D.C. / Covelo, California

A Shearwater Book
published by Island Press

Copyright © 1995 Loren R. Graham

Shearwater Books is a trademark of The Center for Resource
Economics.

Library of Congress Cataloging-in-Publication Data
Graham, Loren R.
A face in the rock : the tale of a Grand Island Chippewa /
Loren R. Graham.
p.   cm.
"A Shearwater book."
Includes bibliographical references and index.
ISBN 1-55963-366-2 (acid-free paper)
1. Ojibwa Indians—Michigan—Grand Island—History.
2. Grand Island (Mich.)—History.   I. Title.
F99.C6G65   1995
977.4'15—dc20   94-46253
                                                              CIP

Printed on recycled, acid-free paper

Manufactured in the United States of America

10 9 8 7 6 5 4 3 2 1

To

Powers of the Air

and

the spirit of his island

CONTENTS

A FACE IN THE ROCK

PROLOGUE

THIS IS A story about the loss of a home, Grand Island in Lake Superior, by a band of Native Americans, members of the Chippewa tribe of the Upper Peninsula of Michigan. It is also the story of the fates of the land and water in that area, and of other creatures that lived there. Both aspects of the story involve death and sorrow, but both also end with rebirth and hope. This story of shadow and light came to me in a personal way, as you will see.

Many Native Americans are properly concerned about their relics, legends, and history being appropriated by others and used for purposes alien to those of native traditions. Although I believe that this book is not such an appropriation, I realize that it may be open to such criticism. I am, after all, a white man, and this story is mainly about a Native American community from the tribe of Chippewa (Ojibway, Anishinabeg) and a specific Native American, Powers of the Air. Yet the account that follows is not solely about Native Americans; it is an intertwining of Native American history with white history, with all the contradictions, violence, prejudices, and reconciliations common to the merging of these histories. Many major characters and the people who told me about them are

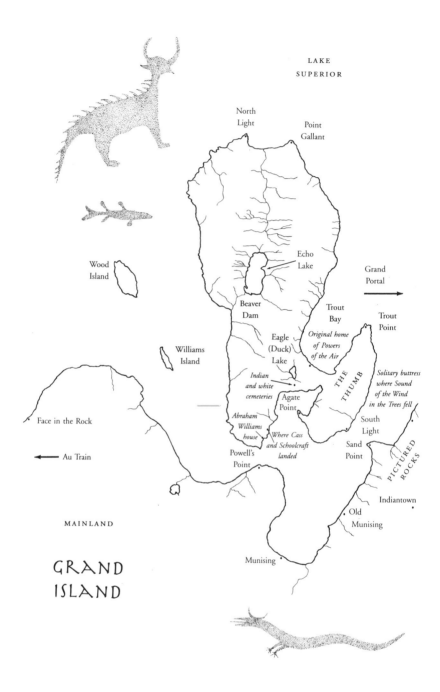

LAKE
SUPERIOR

North
Light

Point
Gallant

Echo
Lake

Wood
Island

Grand
Portal

Beaver
Dam

Trout
Bay

Trout
Point

Williams
Island

Eagle
(Duck)
Lake

*Original home
of Powers
of the Air*

*Indian
and white
cemeteries*

*Solitary buttress
where Sound
of the Wind
in the Trees fell*

Agate
Point

Face in the Rock

T
H
E

T
H
U
M
B

South
Light

Au Train

*Abraham
Williams
house*

*Where Cass
and Schoolcraft
landed*

Sand
Point

P
I
C
T
U
R
E
D

R
O
C
K
S

Powell's
Point

Indiantown

MAINLAND

Old
Munising

GRAND
ISLAND

Munising

of mixed Native American and white descent or are partners in mixed marriages. Henry Schoolcraft and William Cameron, central figures in this narrative, were white men who married Chippewa women. Harry Powell and Julia Cameron, who served as important sources, were of mixed descent. Thus the history told here does not belong entirely to any one group; it is *our* history, one that neither I nor any other person—white or Native American—can possess, but one that we must share.

I am part of this history not through descent but through marriage. My wife's family is deeply connected with the place where this story occurs and with the people in it. Columbus Horatio Hall, my wife's grandfather, met Powers of the Air over a century ago and learned the kernel of this story, and he then passed it on to his descendants who have come back to Grand Island every summer since. Without that oral and written tradition, I would have been unable to reconstitute this history—certainly not in this form.

I first went to Grand Island in 1954 with my fiancée, Patricia Albjerg. Her home there, a cottage on Trout Bay with neither running water nor electricity, had been passed down through the women in the family, including her mother. Patricia herself had summered on the island since infancy. As the only child of Marguerite Hall and Victor Albjerg, it was clear that Patricia would be the next member of her family responsible for the cottage. And as the destined husband of Patricia, I knew that I was on trial. Would I come to share the family passion for this primitive place and its stories?

During most of the next twenty years Patricia, our daughter, Marguerite, and I spent our winters on Manhattan but returned each summer to Grand Island. The two islands were similar in only one respect: Manhattan is twenty-three square miles in area, Grand Island a little over twenty-two. Manhattan at that time had a year-round population of one and a half million people, Grand Island a permanent population of one person (today it is zero). Different as the two islands were, both fed my soul, expanding my sense of the ways in which one could and should live. In Manhattan my daily orbit of home and work encompassed only a few blocks, and I lived

and labored totally oblivious to the weather or the season; I cannot
remember once checking the paper or radio to learn the wind direction. I was happy with my teaching and research and sought diver-
sions not in nature but in dinners and conversations with friends,
in visits to theaters and restaurants. On Grand Island, in contrast,
nature, and especially weather, ruled our lives. Seasonal time was
measured by the week. We learned when to expect, usually within
a few days, the appearance of violets, daisies, wild iris, moccasin
flower, Indian pipe, sweet william, mushrooms, strawberries, raspberries, blueberries, thimbleberries, and blackberries. We foraged
all over the island, sometimes covering twenty miles in a single day.
Wind direction was the most important variable in each day, for it
determined the temperature, and thus hiking and boating possibil-
ities. A swing of the wind from the south to the north, off frigid
Lake Superior, could drop the temperature forty degrees within a
few minutes. A storm on Manhattan went largely unnoticed unless
it soaked me on the way to the library; a storm on Grand Island
overwhelmingly shaped our experience of the place, especially if it
caught us out on the lake in a boat, or if it isolated us for several days
by blowing trees down on the trails and roads.

Since 1978 my wife and I have lived for most of each year in
Cambridge, Massachusetts, a middlescape environment by com-
parison with Manhattan. By the time of that move, however, we
had come to rely on the contradictory nutrients of urban concrete
and pine forests for sustenance. Like migratory birds, we have
thrived in contrasting environments.

During my first years on Grand Island I regarded it only as a re-
treat to nature, and was unmindful of its social history. Then I
came upon the diaries and letters of members of Pat's family and
began reading them in the evenings. They told of a time when a siz-
able Chippewa community had lived on the island. Pat's grandfather, an ordained minister as well as a college professor, had bap-
tized their children, preached in their church, and officiated at their
funerals. But by the early twentieth century few Chippewa remained. What had happened to them? I decided to investigate, and
during the next thirty-five summers I collected information.

In the 1950s and 1960s the authoritative but idiosyncratic sources of information about Grand Island were its only permanent resident, Harry Powell, and its caretaker, John Lezotte. Harry, a descendant of an original island family, lived on the south end in an old house that was a jumble of junk; anything carried into it and put down was immediately lost forever. John, an employee of the Cleveland Cliffs Iron Company, which then owned the island, lived on the mainland most of the year but came to the island every day, and during the summers he lived in a small house there. John and Harry told me that in the early nineteenth century there had been a Chippewa village on Trout Bay.

Harry was the great-grandson of Abraham Williams, the first white settler of the county, and his mother was part Chippewa. He told me that, according to old stories he had heard, "something bad" had happened to the Indians who lived on Trout Bay, and after the calamity they disappeared from Grand Island; but before the group vanished, he said, someone had carved the face of one of the islanders into a cliff near the small village of Au Train, about seven miles away on the mainland. Harry told me how to find the face, and when I asked him whose it was he advised me to go talk to Julia Cameron, a Chippewa woman in Au Train who still spoke "the old Indian language" and whose grandparents had once lived on the island.

One day soon afterward I crossed the strait between the island and the mainland in an old wooden boat shared by four summer families on Trout Bay, then drove my car westward toward Au Train. When I reached the beach, I walked eastward along the shore until I reached a cliff. Starting from a height of about three feet at its farthest point inland, it rose to about twelve or thirteen feet near the shore. Its surface was rough and dark, fissured in many places by small, wavy cracks. Here and there lichens clung to the rock, light green with an occasional splash of orange and a few small dots of red. I moved slowly outward to the shore, where waves raised by a light wind soon soaked my sneaker-clad feet. It was August, but the water was cold. Nowhere did I see a face.

I was disappointed. Harry had fooled another summer person

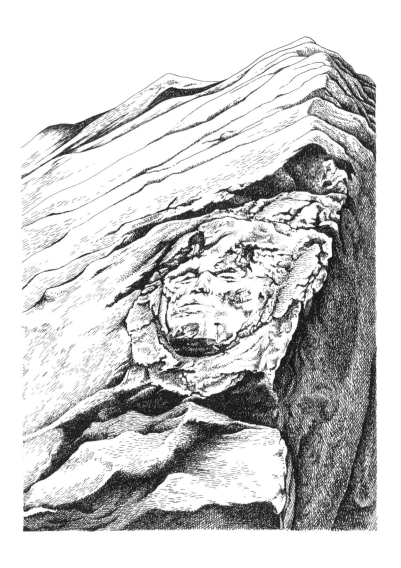

with a tall tale. Still, I stood and searched the cliff, desperately wanting Harry's story to be true. Then, about five feet above my head, I saw it—a little constellation of lines and protuberances that formed a face, slightly larger than life size, blending in with the rock and partly covered by lichens. Was it man-made or nature-made? Looking closely, I saw chisel marks for the individual eye-lashes and another shape—perhaps a high collar or decoration—

extending beyond the right cheek. It was man-made all right, and ancient.

I returned to my car and drove about half a mile to Au Train. At its one business establishment, a combination store and post office, the proprietor directed me to Julia Cameron's house at the far end of the next street. It was a small, unpainted clapboard affair, with a listing gutter above the door. The square logs exposed by gaps in the surface boards told me that the house was a very old one. A middle-aged woman with a sweet face and wide round elfin eyes answered my knock.

"Are you Julia Cameron?" I inquired. "I am Loren Graham, a summer resident of Grand Island. Harry Powell told me your grandmother lived on the island, and he thought you could answer some of my questions about the history of the place."

My simple statement set her off on a monologue that filled my imagination with the richness of the legend of Grand Island.

"Oh lordy, yes. I know lots about my grandmother Sophie Nolan and her husband, William Cameron. He was the keeper of North Lighthouse on Grand Island for many years. They lived up there miles from anybody, all year round. They would go sometimes six months in the winter without seeing another human being. She died in the lighthouse one winter, and he was so sick about it that he mourned her for years. He was a very educated man. He knew six or seven languages—English, Greek, Latin, Chippewa, and some others I have forgotten.

"He got a job with the Lighthouse Service and came to Grand Island in the early 1870s. They came from Sault Ste. Marie when my grandfather was given trouble by other whites for 'going native'—learning the Chippewa language, marrying a Chippewa woman, and living with the local Indians.

"My father, Frank, was raised on Grand Island and grew up speaking Chippewa. He married a Chippewa woman, just like his father. We spoke Chippewa at home, too, but we lived mostly over here on the mainland. Many of the Chippewa didn't live on the island year round. They came in the summers to fish and gather blueberries. A long, long time ago there was a permanent village on the island, over on Trout Bay, but that was before my grandparents' time.

"Have you ever noticed the face on the cliff down on the beach? It's the face of an old Grand Island Indian. Not many people know about that face. There are lots of old stories about it. My father told me it was put there by one of Governor Cass's men. Cass was governor of Michigan territory at the time, and he came through here in the early 1800s with an exploring party, long before any Camerons were here—or anybody else, except the Chippewa. Cass apparently stopped on Grand Island, where they met an Indian who told them all sorts of stories. Later one of Cass's men carved the Indian's face in the rock.

"The original Grand Island Indians were a little strange, so there were many stories about them. They didn't want to have much to do with the rest of the Chippewa. Seems they preferred just to be left alone on the island. But after a while they got into some sort of fight on the mainland, and eventually moved their village off the island and disappeared. From that time on, no one had any special claim to the island. Instead there was a lot of movement on and off it by the mainland Chippewa, who would come in the summers mostly to hunt and fish. By the end of the nineteenth century the surrounding towns—Munising, Seney, Newberry, Grand Marais, and others—grew much larger than the little village on Grand Island, and so the island became just a backwater, with not many people living on it. Eventually everybody moved off."

My conversations with Harry Powell and Julia Cameron over thirty years ago left me intrigued with the mystery of what had happened to the original Grand Island Chippewa, what sort of calamity or fight had occurred, and the identity of the face on the rock. I collected information from many different directions, and all of it I placed in file folders I kept on the island.

One of the most important sources turned out to be the members of my wife's family. Pat's grandfather Columbus Hall left several volumes of diaries and numerous letters written on Grand Island. He met and talked to the aged William Cameron, and he engaged Cameron's son Frank (Julia's father) as a guide. Pat's grandfather died before she was born, but several of his children were still very much alive when I became interested in the history of the island. His daughter Letitia Hall Carter, my wife's aunt, came to the island almost every summer from 1888 to 1969, and loved to tell stories about the island. I talked to other descendants of old Grand Island families, including Lewis DesJardins and Aaron Powell. I visited Julia Cameron several more times before her death in 1994 and also had many more conversations about the island with John Lezotte and his brother Tony, a tugboat captain and a former mayor of Munising. And I looked up the records of Lewis Cass's 1820 expedition to Grand Island. Even more revealing were the papers, books, and articles of Henry Schoolcraft, who accompanied Cass on that expedition and who visited the island several times in the early nineteenth century. Schoolcraft married the granddaughter of a Lake Superior chief and wrote voluminously on the history of the area. In 1972 Patricia and I and our daughter, Marguerite, purchased and restored the abandoned North Lighthouse on the island, where Julia Cameron's grandparents had lived, and in subsequent years I located William Cameron's diaries, including his poetry. Many Sault Ste. Marie Chippewa helped me to find still more information. Eventually I solved the twofold mystery: I learned what had happened to the original Grand Island Chippewa, and I found out whose face was on the rock near Au Train. In the process I gradually reconstructed the pieces of a long story that I now call the Grand Island Legend. I found that the legend was connected to broad themes in American history and literature in ways that I never could have imagined.

Although I garnered a great deal of information, I make no claim that everything recounted in the following pages is authentic history. All references to and quotations from known historical figures (such as Jacques Marquette, Lewis Cass, Henry Schoolcraft, John Calhoun, Henry Wadsworth Longfellow, Louis Agassiz, Gifford Pinchot, Stephen Mather, William Mather) are accurate, taken either from their papers or from established sources. But in addition to much truth, the story also contains fiction. In my usual work as a professional historian, when I encounter several different versions of the same event I expend great effort to decide which is closest to the truth. In this case, though, when I have encountered several different versions of the same event, I have unabashedly selected the most interesting one, so long as it did not fly in the face of what I know to be true. Furthermore, I have filled in gaps in the story with my own imagination, relying on my historical sense of what *may* have happened and inventing dialogues and descriptions. The final product is definitely not history in the academic sense, but neither is it exactly fiction. As an account that can be taken as historical but is not entirely verifiable, I prefer to call it a legend. Or perhaps it should be called "imaginative history."

Spellings of words in the Chippewa or Ojibway language vary greatly. Whenever possible, I have used the spelling given in Frederic Baraga, *A Dictionary of the Ojibway Language* (1878; reprint, St. Paul: Minnesota Historical Society Press, 1992). For the Chippewa terms, especially proper names, that have come to me from local sources and have no equivalents in Baraga, I have used current local spellings.

Whether to use the terms "Indians," "Native Americans," or "native people" is a problem I have solved pragmatically: when I am speaking independently as an author I usually say "Native Americans" or "native people," but when I refer to people speaking in the past in the context of the legend related here I follow the custom of that time by using the term "Indians."

L. G.
Grand Island

THE IDYLLIC LIFE

THE CHIPPEWA CALL Grand Island Kitchi-miniss, or Great Island, and those living there called themselves *minissing-endanajig*, islanders. In the seventeenth century the first white explorers of the area, Jesuit priests from France, gave the French equivalents of Native American names to many of the places they visited. On a crude map of Lake Superior drawn in 1673 by Father Jacques Marquette the only islands of the lake shown with a name are Les Grandes Isles, the plural embracing what now is called Grand Island and the nearby smaller Wood and Williams islands. One of the explorers recorded the Chippewa name of the largest bay on the island as Namegossikan, Place Where There Are Trout; today it is known as Trout Bay. Similarly, the nearby small town is called Munising, an adaptation of Kitchi-minissing, At the Place of the Great Island.

The fifty or sixty Chippewa who lived on Grand Island in the eighteenth and early nineteenth centuries, before any white people had moved into the area, had no trouble finding adequate food. Fish were so plentiful that in the deep, clear water they appeared as shoals of silvery metal extending along the colored cliffs and brown beaches that ringed the island. The island Chippewa caught the fish by going out in their canoes at night and holding flaming torches over the water; attracted by the light, whitefish and trout would swim to the surface near the canoes, and the Chippewa would spear and haul them in. Smoked or dried, the fish could be kept for months. The island also provided many other kinds of food: deer, moose, and bear and, during certain seasons, ducks and geese. In the summer strawberries, raspberries, and blueberries were abundant. The blueberries on Trout Bay were so numerous that, dried and preserved, they could be consumed all year round.

In the center of the island was a large lake, a mile long and almost half a mile wide. The Chippewa felt that there was something mysterious and unnatural about the lake. Its southern shore, where it debouched through a swamp into Lake Superior, was the widest and straightest part, stretching hundreds of feet in a way that suggested manufacture. Somebody or something had made that shore with its hands and feet. Perhaps it was a forest spirit, or perhaps it was a society, working for a common goal. Actually, it was both. For centuries the lake had been the home of an immense beaver colony, perhaps the largest in Lake Superior, and the beaver had built a dam extending fifteen hundred feet along the southern edge. Gradually, as the dam metamorphosed into an earthen shore, with trees and bushes growing upon it, the beaver moved their construction sites farther south, building additional smaller dams.

The island Chippewa called this large lake Puswawa-sagaigan, Lake That Resounds, because when they stood on the ancient dam at the south end and shouted to the north, after a few seconds the beaver spirit, *amik-manito*, would throw their voices back. Many years later Echo Lake would be featured in limnology textbooks as one of the largest "social lakes" (formed by a society of nonhuman animals) in the world. In the 1980s palynologists established through the presence of pollen deposits that the beaver dam at Echo Lake was approximately seven hundred years old.

For the Grand Island Chippewa this beaver colony was a source of great fortune, providing for their existence in many ways. Beaver meat was excellent, and beaver tail, *amikosow*, a delicacy sought everywhere. Even more important, beaver furs made life during the cold winters endurable. Beaver furs were used as robes *(muttataus)* on frigid days, as rugs in the lodges, and as insulation on walls.

The lodges of the island Chippewa were constructed according to a standard procedure. Tall young saplings, usually of birch, maple, or beech, were cut and thrust vertically into the ground to form an oval. Then the saplings were bent over and fastened together in pairs with basswood cord. Young trees were tied horizontally to the arches to form crossbars. Finally the lodge, now resembling a large basket, was covered with overlapping rows of birchbark *(apakwei)* sewn together with split spruce. This arrangement ensured that rain would run off the walls instead of through them. A mat woven of cattail stalks served as a doorway, with heavy stones placed on the bottom edge to keep the wind from blowing it open. The lodges were left thus throughout the summer. With the advent of cold weather, several layers of beaver fur interspersed with cattail mats were added to the outer walls and covered by a layer of slippery-elm bark. With cedar boughs and beaver rugs on the floor, and robes to sleep in around the central fire, vented through a hole in the roof, the Chippewa could remain warm even when their lodges were blanketed by snow. This mode of life required a great number of beaver pelts, but the Echo Lake colony was so large, and the Grand Island Chippewa so few, that the beaver population was not threatened. The two societies lived in balance.

The main Chippewa village was located near the south end of the island, amid large pine trees on the stretch of sandy land extending from the main part of the island to the Thumb (Kitchi-onindjima). This place, too, had its own spirit. According to Chippewa tradition, the Thumb had once been separate, but many centuries before, one of the forest spirits had taken pity on the lonely smaller island and joined it by a sandy isthmus to the larger one. Geologists would later confirm this history in its broadest outlines and describe the sandy stretch as a tombolo, a formation with distinctive characteristics and vegetation. Fresh water lay only two or three feet below the surface, and the sand of the Thumb was covered with interconnected islands of lichens and moss, which the Chippewa gathered and placed under their beaver rugs to help them to sleep without bad dreams.

Not far from the village was another small, shallow lake, which the Chippewa called Migisi-sagaigan, or Eagle Lake. The imposing birds would gather there when they wanted easy catches of bluegills and small pike, snatching them when they came too close to the surface. The name Eagle Lake persisted through the last decades of the nineteenth century. Early in the twentieth century it became known as Duck Lake after an industrial tycoon from Cleveland purchased the island and converted it into a private game preserve.

Between the small lake that enticed the eagles and the large inlet, now called Murray Bay, which separated the island from the mainland, stretched about a thousand feet of land. To the island Chippewa this was a sacred place, to be visited only on special, and mournful, occasions. It was their burial ground. On the death of one of their band, close relatives washed the body, dressed it in the person's best clothing, and painted a round brown spot upon each cheek and a horizontal line of vermilion extending from the brown spots to the ears. The body was placed in a low peaked wooden structure, or spirit-house, with the feet pointed to the west, the direction of the spirit's journey. Markers placed at the grave indicated the totem of the family. For four days after burial, fires were kept burning nearby and food was placed on a ledge at a small hole in the western end of the spirit-house where the spirit could emerge.

The Chippewa burial ground was located among great white pines that to this day have never been logged. In later years Abraham Williams, the earliest white settler, and his family and descendants were buried not far away, a little closer to the great lake. People who live in the area and go there by boat for picnics still refer to the spot as Cemetery Beach.

In their hunting, eating, and festivities the Grand Island Chippewa took advantage of all the island's peculiarities. One of these was the rock formations that underlay the island and engirdled most of its coastline. The Chippewa often spoke of the island as a place "where the soft rock meets the hard rock." Their observation would later be supported by geologists. On the mainland to the east of the island the rock is primarily sandstone, which the Chippewa called *pingwabik;* on this shore in sight of the island are great cliffs, stretching three hundred or more feet high, which the Chippewa called the Painted Rocks and today are known as the Pictured Rocks and are now a national park. These sandstone cliffs were easily chiseled by water and wind into a great variety of shapes. To the west of the island the sandstone gradually disappears, to be replaced by crystalline rocks in which white settlers frequently found ore. The Marquette Iron Range, forty miles to the west, and the Mesabi Range, much farther in the same direction, would become the sources of the iron ore for much of America's steel. The harder rocks to the west were also rich sources of copper and even of gold; in the late twentieth century Marquette is the site of the only gold mine in the United States east of the Mississippi.

At Grand Island three different rock formations meet, two of them Cambrian sandstones of varying hardness (the Jacobsville and Munising formations) and one of them crystalline rock of the Ordovician period. This convergence produced remarkably parallel rows of red and white rock, extending from the shore of the island into the lake. The Chippewa were convinced that these lines, like the big lake in the center of the island, were of supernatural origin: one of the forest or water spirits, a *manito*, had created Grand Island in a fit of playfulness. Geologists have explained them as "clastic dikes," the results of an opening up of a set of shear joints in the limestone of the Jacobsville formation and the injection under hydrostatic pressure of a slurry of limestone from the younger Munising formation.

But not all the designs on the island were rectilinear. Midway up the western side, cliffs rising almost four hundred feet above the lake formed gray corduroylike layers of differing hardness. Farther north these yielded to red and white twists and sworls that recalled heavenly, human, and animal shapes. Here one could see the moon, there a bear, and even an "evil eye" demonically staring at trespassers.

For the Chippewa the meeting of rocks of different hardness, working against one another in such wondrous ways, produced useful as well as entertaining results. Along the rocky shores of the island several sandstone shelves or tables extended into the lake, standing two or three feet above its level when it was calm. Over the years waves formed shallow depressions in these flat shelves. Whenever a small piece of granite or another hard material was washed into a depression or fell into it from the cliffs above, the action of the waves during a storm would cause the little hard rock to scour the soft walls of the dimple, wearing away the stone like a pestle in the mortar of sandstone. Eventually these small holes were enlarged to as much as three feet deep and a foot or two across.

The Chippewa used these holes as cooking pots. Before the arrival of white traders, such items were difficult to come by. Even these Grand Island pots were by no means convenient, since there was no way to build a fire under them and they were located on the northern shores, far from the village on Trout Bay, and accessible only by canoe. The Grand Islanders contended with these problems by heating the pots in an unusual way and by using them only on special occasions.

STONE POT

On these feast days everyone from the village would gather on the rocky shelves, clean the pots of debris such as leaves and sticks, and fill them with water carried from the nearby lake in birch containers called *watabi-makak*. Then they would build a bonfire nearby and place many hard blue-black stones, each about the size of a man's hand and brought there specifically for the purpose, among the red-hot coals. When the stones glowed red-hot, the cooks quickly pushed them into the pot, usually with deer antlers. When the stones had dispersed their heat into the water and cooled, the Chippewa retrieved them with the antlers or with birch scoops and returned them to the fire. By repeating this activity many times, the islanders could bring an entire pot, or several pots, to a boil.

Into the hot pot the Chippewa placed whitefish, potatoes, corn, mushrooms, herbs, and, best of all, their two delicacies: tender cheeks of giant lake trout and small pieces of beaver tail. The result was a bouillabaisse that, until the iron pots of the white man's civilization arrived, only the Grand Island Chippewa could produce in enough quantity to feed a large crowd. They regarded these Stone Pot Feasts as an important part of their culture. The only outsiders who participated did so by special invitation, an honor extended by the island band as a whole.

Native Americans had lived along the southern shores of Lake Superior, and on Grand Island itself, for centuries, and they left scores of legends behind. Almost every promontory on the cliffs and every bay along the shores had at least one legend connected with it. Several of the legends centered on Grand Island itself. The favorite legend of the island Chippewa was about Mishosha, the magician of the lake. Many years earlier Mishosha had made Grand Island his home. He possessed a beautiful canoe that, if the proper magical words were spoken as one slapped its birchbark side, would sail so rapidly through the lake that within a few minutes one could go anywhere in Lake Superior. Once Mishosha had uttered the secret words, he could direct the canoe merely with his thoughts. The wish to go to a certain island on the lake would bring him to that very spot within a few minutes. Mishosha used this great power to rule all the islands of the lake, extending his realm from his home on Grand Island. Each of the islands had its own attractions and treasures, and Mishosha reaped the benefit of them all. In a few seconds he would go to one of the small islands west of Grand Island, now called Williams Island, where he would gather seagulls' eggs, a great delicacy, from the large rookery that had always been there. Or he would order his canoe to go to distant Michipicoten Island, a mysterious and isolated place with eagles so large that a person could ride on their backs. Or Mishosha would direct his canoe to the smaller Adikiminis island, now known as Caribou Island, which was at the center of the lake, hence remote and rarely visited. Adikiminis was surrounded by sands that were said to contain much gold. It was sometimes called the Island of Yellow Sands, and its treasure was protected by enormous snakes. Mishosha could even travel in a few minutes to the farthest and largest island of all, Minong (actually an archipelago now called Isle Royale), where there were sturgeons so large they could swallow a human being in one gulp. On Minong he obtained bits of soft copper. Once returned to Grand Island, Mishosha would direct his canoe along the shoreline and little bays where in secret spots he gathered beautiful stones and pebbles—agates, carnelians, hornstone jasper—that he used to make fine jewelry, placing the stones in the soft copper from Isle Royale.

For many years Mishosha employed the great powers of his wondrous canoe in a benevolent way. Throughout the Lake Superior region Grand Island was known as Enchanted Island because it was the home of the magician of the lake. But as Mishosha grew older he began to take his power for granted and to believe that he was omnipotent, becoming selfish and vindictive. He would capture his enemies and take them in his charmed canoe to some distant island to be devoured by wild beasts. On Michipicoten the victims would be killed by giant eagles, on Adikiminis by fierce snakes, and on Minong by enormous sturgeons. Eventually he kidnapped a young man named Panigwun (Last Wing Feather), who was determined to overthrow him. Each time Mishosha took Panigwun to a far-off island where a beast was supposed to kill him, Panigwun would outsmart Mishosha with the aid of sympathetic wood and lake spirits, who had grown tired of Mishosha. Eventually Panigwun learned from Mishosha's daughter the magic words that controlled the canoe. Panigwun wrested the canoe from Mishosha, and defeated him in winter combat, on the snow and ice. Mishosha froze to death; the feathers in his headdress turned into leaves, and his legs grew downward into roots. Mishosha became a giant birch tree on the edge of Grand Island, leaning toward the water.

All the Grand Islanders considered Mishosha and his magic canoe a part of their history. They even knew the secret words, *chemaun poll*, that Mishosha uttered to make his canoe fly over the water, but they would not reveal the words to any outsiders. Sometimes when Grand Islanders were out on the lake and a storm endangered their return, they would slap the side of their canoe and exclaim *chemaun poll*. The Grand Islanders admitted that at such moments their canoe would not fly with the speed that Mishosha had been able to summon, but they were convinced that it went faster than before.

Life on Grand Island for the Chippewa in the eighteenth and early nineteenth centuries also presented occasional dangers and hardships. Injuries during hunts, especially bear hunts, were common. Fishing in the great lake was treacherous; sudden shifts in the wind could blow light canoes far out into the lake or capsize them. The bodies of the drowned were almost never recovered. The Chippewa said of their lake, Kitchigami (Great Lake), what white people say today: "Lake Superior never gives up its dead." The waters in the depths of the lake were so cold, varying no more than five degrees from winter to summer, that bodies did not putrefy and rise to the surface from the gases of decay, but remained on the bottom in a deep freeze. The Chippewa believed that a person to whom this happened was doubly condemned, since without a proper burial the spirit could not escape.

The winters, despite the warm lodges, were a special trial. Famine was almost unheard of, but sickness was common. Old age was

MISHOSHA

particularly cruel. The elderly usually could not see well, their bones ached from the cold, and they were dependent on their children for food and care.

The Grand Island Chippewa were free from one affliction that visited most other Native Americans. No one could remember when a Grand Island Chippewa had fallen in battle, nor could anybody recall when one of them had taken the scalp of a rival group or tribe. Secure on their island, they did not engage in war. Perhaps because they did not have war heroes, the Grand Islanders placed special emphasis on the sport of running. Speed of foot was highly honored among the islanders. All the young men, and many of the young women, practiced running on the island's beaches, and the best of them were celebrated by all the other Chippewa.

Running was not only a sport but also a means of hunting, called "running down the game." In the winter, groups of the island Chippewa would drive the deer in a small herd by pursuing them on snowshoes while the deer struggled to get through the deep drifts. Since carrying a dead deer back to the village on snowshoes was difficult, the Chippewa had a strategy to make the task easier. They would chase the deer until one was exhausted. Then they would wait for it to recover a little, while getting their breaths themselves, then resume the chase, each time directing the deer toward the village. When the hapless animal was a short distance from the Chippewa lodges, the hunters would close in on it and kill it, usually with no larger weapon than a stone knife or a small spear.

Late in the second decade of the nineteenth century, before any permanent white settlers had come to the area, and the moment when the Grand Island Legend begins, the fastest runner on the island was a boy approaching manhood named Pangijishib, or Little Duck. His father, Pikwakoshib, or Autumn Duck, was the chief of the Grand Island Chippewa. His mother was Medweackwe, or Sound of Wind in the Trees.

Early each morning Little Duck would go for a run on Trout Bay beach, which extends for a mile between two great arms of cliffs to the east and the west. For him this run was as much an emotional as a physical experience, sustaining him through the rest of the day.

He sped along the line where the water met the sand, splashing his bare feet noisily. Once he gained speed, he would tilt his head back and look at the sky, and try to make the splashes occur closer and closer together. His parents and other villagers would stand and watch, marveling at his speed. Occasionally another Chippewa would try to pace Little Duck but would always fall far behind.

Little Duck always ran his first lap to the west. Every morning, about two-thirds of the way, as he came abreast of a certain pine tree standing back from the beach, he would lower his head and look to the northeast across the bay toward the point at the north end of the Thumb, until a large promontory on the mainland suddenly emerged. This rock, standing three hundred feet high, was a great arch later known as Grand Portal. It was a part of the long stretch of colorful cliffs and ramparts known today as the Pictured Rocks. The arch of Grand Portal was so high that photographs taken in the last part of the nineteenth century show two- and three-masted schooners sailing right through it.

Little Duck would run his three or four laps back and forth along Trout Bay beach, take a quick dip in the lake, and then return to his lodge. It was a ritual that everyone in the village knew. Other young men and women also often ran along the beach, but they usually did so at a different time, when their speed would not be compared with that of Little Duck.

FROM PEACE
TO WAR

WHILE THE GRAND Island Chippewa stayed on their island, their life was likely to be peaceful and secure. But when they journeyed to the mainland to visit with their fellow Chippewa, they fell into a disagreement that intensified over the years. This rift festered even though several of the same *do-daim*, or clans, existed on both the island and the mainland. The disagreement was about warfare. The Chippewa occupied the shores surrounding Lake Superior. Their traditional enemies, the Sioux, lived farther west, in present-day Minnesota and Iowa. Warfare between the tribes was almost constant, and the enmity between them was so deep that members of other tribes would say, "The Chippewa and the Sioux are the reddest of all the redmen, red with each other's blood."

Although warfare had always been a part of mainland Chippewa life, it had been greatly exacerbated by pressure from whites moving in from the east and pushing the Iroquois westward from their homes in present-day New York State. The high point in the Iroquois drive to the west occurred in 1662, when a group of them were annihilated by the Chippewa on the south shore of Lake Superior, about eighty miles east of Grand Island. The spot is known today as Point Iroquois, and local Chippewa still think it is haunted because of the Iroquois bones thought to lie beneath the sand.

Although by the early nineteenth century only the Sioux remained a serious concern, the Chippewa still considered themselves to be caught between two enemies, and they continued to push the Sioux back. Although the Sioux were becoming plains Indians as they were thrust westward, they still dreamed of taking back their forest lands between Lake Michigan and Lake Superior, and they regularly sent war parties against the Chippewa. The Chippewa, in turn, regarded the Sioux as vengeful warriors who would destroy them if they did not continue to push forward.

The Grand Island band of Chippewa, alone among their tribe, had never joined the war parties that traveled westward to fight the Sioux. Feeling secure against invasion, on their island, they saw no reason to fight any other people. All were the children of the spirits of the land and water. The Sioux and the Chippewa, they thought, had their own territories, with space for their activities and food for their stomachs, and should live together peaceably. For themselves, they wished only to enjoy their beloved island with their children and leave the rest of the world alone.

But the rest of the world would not leave the Grand Island Chippewa alone. Chippewa up and down the shores of Lake Superior called them cowards, and some began to insist that the Grand Islanders were not really Chippewa, but an alien group, since all Chippewa men were warriors who defended the tribe and increased its power. The Grand Islanders just laughed at this rumor, insisted that they were as good Chippewa as any others along the great lake, and went on with their lives as before.

Contempt for the Grand Islanders became so deep that whenever they went to the mainland, other Chippewa would smirk at them and call them *waubosog*, rabbits. When they saw a Grand Islander they would laugh and place their fingers on each side of their heads, imitating long ears, and hop. Stung by this treatment, the islanders began to discuss among themselves whether their peaceful way of life was correct. Each time they decided that they were living by rules that fitted them, and they continued to refuse to join the war parties.

Only once in recent memory had the unanimity of the Grand Island Chippewa been disrupted. Near the end of the eighteenth century one particularly headstrong young man, Shingwauk (Little Pine), had been so humiliated during a visit to the mainland that he had left his home on the island, joined a war party against the Sioux, and quickly won glory for his exploits. He never returned home. He became a famous chief of a band of mainland Chippewa, then moved to Boweting (Sault Ste. Marie), where he allied himself with the British in their struggle against the Americans for control of the upper Great Lakes. He often dressed in the red coat of a British officer.

Shingwauk was known not only for his military victories but also for his magical powers as a shaman and as a leader of the Midewiwin, or Grand Medicine Society. The Chippewa said that he could transform himself into other living beings. By turning himself into an insect he could travel on the backs of birds at great speed, thereby confounding his enemies by his sudden appearances at widely distant locations. Some Chippewa today, including his descendants among the Pine family north of Sault Ste. Marie, claim that he was the last of the artists of the pictographs painted on the rocks at secret spots along the shores of Lake Superior. In addition to being a noted warrior and artist, Shingwauk is said to have collected a large library of written and painted birchbark books on Chippewa traditions. Shingwauk was an intensely proud man, difficult toward all strangers, alienated from his peace-loving relatives on Grand Island, and often hostile to white men. In 1855 the German ethnographer Johann Georg Kohl went to visit Shingwauk with the intention of examining his birchbark library, only to find that Shingwauk had recently died and had "destroyed all his papers and birch-barks, and painted dreams, dances, and songs, shortly before his death."

One result of the friction between the Grand Islanders and the mainlanders over the issue of warfare was that the islanders left

SHINGWAUK

their home less and less frequently. They could find what they needed on the island, they were among friends there, and they avoided difficulties. They heard none of the talk of war so common around the lodge fires on the mainland.

The relative isolation of the Grand Islanders only intensified the distinctiveness of their way of thinking and their way of life. While the mainlanders frequently traveled to the trading post at Sault Ste. Marie, the Grand Islanders rarely made the trip. The mainlanders, unlike the Grand Islanders, acquired more and more guns. Guns were very useful for a battle against the Sioux. The Chippewa on the mainland liked to think that the reason for their success in pushing the Sioux back was that they were better warriors, but another important reason was that, being east of the Sioux and in closer contact with whites, they had more guns. Guns were less useful for daily hunting, since powder and shot also had to be acquired from the whites and were expensive. Only Autumn Duck, the chief, and one other Grand Islander, Kitchigwanan, Large Feather, had guns, but they rarely had ammunition.

Autumn Duck was the best hunter with bow and spear on the island, and he took great pride in bringing back game in the traditional way. On the one or two occasions a year when the Grand Island men would bring home a bear, killed by arrows and spears, all the islanders would celebrate. Then Autumn Duck would pay homage to the bear, calling the dead animal *Makwa, nindangoshe* (Bear, my cousin), begging its forgiveness for what they had done, and thanking it for providing sustenance.

For a long time the islanders and the mainlanders went their somewhat separate ways. Occasionally the mainlanders would invite the islanders to join them on a war party to the west, but the islanders always declined, saying that they saw no need for war.

In the first years of the nineteenth century, relations between the Chippewa and the Sioux became even more inflamed. Emotions on both sides built higher and higher. The enmity between the two tribes became so deep that communication was nearly impossible. Any Chippewa finding any Sioux in the woods immediately attacked, and vice versa. In order to exchange threats they wrote "letters" to each other on pieces of birchbark held flat by two sticks along the sides, each epistle about fifteen inches wide and eighteen inches long. On this bark the warring tribes wrote pictographs indicating the frightful things they intended to do to each other. They placed these letters on the borderland areas between the two tribes at places where each knew the other would find them. The letters were prominently displayed on saplings cut to head height and slit at the top to hold the letters. Often the same sapling was used by both sides as a mailbox. Curiously, neither side would wait at the mailbox to ambush the other. There was a tacit understanding that this communication-at-a-distance should be allowed to function undisturbed. But in the letters both the Chippewa and the Sioux promised soon to send war parties against the other and to show no mercy during the encounter.

As before, the Grand Island Chippewa stood apart from all this war preparation and continued their island life. Finally, though, just before the mainland Chippewa prepared to meet the Sioux in a battle, they sent a delegation to Grand Island with an unusually tough message. A chief from the western end of Lake Superior said to the Grand Islanders, "This is the final test of whether you are really members of the Chippewa tribe. Either you join us in war against the Sioux or we will declare to all the Chippewa throughout our territory that you are not only cowards but outcasts. If you do not join us, once the battle with the Sioux is over the other Chippewa will consider the Grand Islanders to be enemies. They will move against you and take your island from you."

The Grand Islanders asked for a day to consider this ultimatum. Autumn Duck assembled a council of all the men in fighting shape, a total of twelve, to draw up a reply. One of the men believed that the mainland chief was bluffing. "He is from far away, several days' travel by canoe to the west," he said. "He does not know us as do the Chippewa on the mainland near us. They are almost in sight of

PICTOGRAPH LETTER

us; some of them are our relatives. They may make fun of us, but they know that we are Chippewa, and they would never attack us. We have no tradition of making war. We have few guns. We should remain as we are." But the others thought the situation was more serious. "It is true," they said, "that our nearest neighbors are not likely to attack us. But the issue of our refusal to fight wars has now become known throughout the Chippewa lands, and some of the groups to the south, and especially those to the west, near the Sioux territory, know only one thing about us, that we will not help them against their enemies. They are more numerous than our relatives nearby, and eventually they will attack. And the fact that we do not have guns is not so serious if we go against the Sioux. Few of the Sioux have them either. We will be on more equal terms with the Sioux than we are with our own Chippewa brethren."

The latter argument prevailed. The Grand Islanders decided that there was no alternative to joining the Chippewa war parties. They gave their answer to the mainland chief and began making preparations for war.

Little Duck approached his father and asked to be made a member of the war party. Autumn Duck considered him to be still a boy, but Little Duck pointed out that he was taller than his father and the fastest runner on the island. They might need a courier before or during the battle, and he was the logical person for this role. Eventually Autumn Duck relented, and Little Duck joined the band of warriors collecting weapons and gear.

It had been so long since any Grand Island Chippewa had engaged in war that they had lost the knowledge of how to prepare. One thing they did know: warriors were supposed to look fierce— the fiercer the better. It would help the Grand Island Chippewa to overcome their reputation as cowards if their war paint and war dress were more frightening and daunting than any among the large group of Chippewa preparing to move west. And here the Grand Islanders had an advantage because they lived near the Painted Rocks. To obtain a full palette of strong colors they would take their canoes there.

They crossed to a wet wall of rock rising several hundred feet out of the lake, with vivid streaks of pigmentation. In some places the colors emerged so strongly that it seemed as if large pipes filled with alabaster, jet-black, or ruby dye were dumping their contents directly onto the rocks and into the lake. In other places wavy copaline curtains of copper, umber, or garnet dyestuffs oozed from horizontal cracks. The hued escarpment stretched for several miles, and in the late-afternoon sun it presented a chromatic prospect of variegated tints: ocher, calcimine, argent, verdigris, ivory, black, slate, mahogany, vermilion. Geologists would later explain the colors as surface stains from mineral-laden waters coming down from above through the rock layers. Today thousands of people come in tourist boats every summer to view the patterns. To the Grand Island Chippewa this spectrum of colors was evidence of the special favor their home enjoyed among the spirits of the great lake. They hoped that the same spirits would help them as they prepared to go to war.

Gouging their fingers into wet spots in the seams of the rocks, the Grand Islanders dug out gobs of red, white, black, copper, and green pigment, which they stored in small birchbark containers. Many of the richest deposits of dye were high above their heads. Whenever they could find spots where they could wade along at the base of the cliff, they stood on the rock shelves at the water's edge, and their friends would climb on their shoulders to reach higher. At other locations, several of the islanders managed to climb to the very top of the cliff and then with a rope made of woven vines descended to the most vivid fonts of color. Returning to their canoes, they added their gleanings to the store in the little baskets. They knew that by adding some water to the brightly colored gobs they could preserve the war paint indefinitely.

Upon their return to the island, the members of the canoe party found that the mainland Chippewa nearby had sent a messenger with tobacco to the Grand Islanders, inviting them to a preparatory war dance. Although such invitations had been received before, never had one been accepted. This time the Grand Islanders sent back a gift of their own, an arrow painted red, a certain sign they were going to join the war party.

In preparation for the war dance, the island women helped the warriors with their paint and dress. Not knowing the paint patterns traditionally used by Chippewa going to war, they invented new ones, all designed to make their husbands, brothers, and sons the most frightening creatures anyone had ever seen. The oldest man in the tribe, Nahbenayash (Line of Thunder Clouds), and the keeper of their traditions, helped by telling the women about the pantheon of awful monsters in Chippewa mythology. The women transformed these descriptions into terror-inspiring grimaces and twisted visages in the brightest colors any Chippewa had ever seen. At first the men were embarrassed by their fierce appearance, which seemed so at odds with the amicable customs of the Grand Islanders. Then the decoration became a game. Finally, amusement was replaced by fear as the men saw their friends become deformed specters and they realized that they, too, looked that way. The women urged the men to overcome their fright and memorize the patterns so that they would be able to reproduce them before going to battle against the Sioux.

Once their preparations were complete, the Grand Islanders paddled across the channel. When the mainland Chippewa saw them emerge from their canoes and walk toward them, their hearts were filled with dread and awe. The rabbits from Grand Island had become terrifying brutes. The ground seemed to tremble as the mainlanders saw all the nightmares and horrific legends of the Chippewa coming at them. There was Kwasind, the fearful Strong Man, who could throw rocks across Lake Superior; here was Boshkwadosh, the Mastodon, who crushed twenty hunters at one rush; next to him was the Toad-Man, so ugly that he could kill with a glance; following was Mishemokwa, the gigantic grizzly bear who chomped Chippewa heads like blueberries. And there was Bokwewa, the terrifying humpback magician who brought lightning bolts down on his enemies. All the Chippewa had heard about these figures late at night around the fires, and they had trembled at the thought of ever encountering one of them. Now they were all approaching at once.

The islanders convinced the mainlanders that they were the same Grand Islanders as always, except that now they were willing

ISLANDER DRESSED FOR WAR

to go to war. Any doubts the mainlanders had had that the meek islanders would be an asset to the war party were gone. The Chippewa had a secret weapon, a small band of horrible creatures who could sweep away enemies merely by looking at them.

The mainlanders invited their terrifying new allies to join them in a war dance. Autumn Duck confessed that the islanders did not know exactly how a war dance was done. They had no experience. The mainlanders replied that they would happily teach the islanders to do a war dance if the islanders would help them with their war paint. Both sides enthusiastically agreed to the bargain.

THE BATTLE OF
THE CAVERN

SEVERAL DAYS LATER the war party paddled westward along the southern shore of Lake Superior, their war paint and war dress carefully packed away. Five Grand Island canoes carried two warriors each, and one, slightly larger and belonging to the chief, held three. Autumn Duck sat in the stern, Little Duck in the middle, and Large Feather in the bow. Autumn Duck's craft led the Grand Island canoes, but ahead of them were about twenty others holding the mainland Chippewa with whom they had danced. As they continued westward they were steadily joined by other canoes and warriors. The flotilla was growing.

The war party stayed close to the shore so they could take shelter if the wind intensified. At the end of the second day, near Huron Women's Islands, the present-day Huron Islands Wildlife Sanctuary, they had to make a choice: either stick close to the shore and go deep into Huron Bay, or cut across through open water. They took the more hazardous route but encountered no difficulty. Immediately thereafter they reached an even larger and deeper bay, known today as Keweenaw Bay, and faced the same problem. This time they kept the canoes close to the shore until they reached a promontory known as Pikwaming (today Pequaming), the Hump. Like the Thumb on Grand Island, it had once been an island but had become connected to the mainland by a long sandy isthmus. The Grand Islanders took pleasure in the resemblances of the beach to their own on Trout Bay. Here the warriors decided to embark on a riskier route and cross the open water to the Keweenaw Peninsula, a long finger of land projecting seventy miles into Lake Superior. Once more, the trip was accomplished without mishap; the Chippewa began to feel that their expedition was blessed with good fortune and success.

For centuries the Native Americans and, later, the French explorers who traversed Lake Superior had used a portage bisecting the Keweenaw Peninsula, and the Chippewa warriors followed the traditional route. The first twenty-five miles of this shortcut were a pleasurable journey along present-day Portage Lake, a body of water in many places more similar to a broad river than a lake, but one without current. It passes between picturesque hills on which today are located the cities of Houghton and Hancock, Michigan.

The last few miles of the shortcut, however, were a section that all travelers dreaded. The lake came to an end where a small and twisting stream emptied into it. At first just wide enough for a canoe, the stream became narrower and narrower and was filled with more and more obstacles, such as fallen trees. Finally it disappeared entirely into a deep and stagnant swamp of tamarack and hemlock trees. Here the Chippewa had to drag their canoes through knee-deep mud, stumbling over buried branches while leeches clung to their skin. The U.S. Army Corps of Engineers would later dig a canal through this portage to Lake Superior. The swamp today, dredged out and widened, is known as the Lily Pond. The final section of the portage, where the Chippewa had to carry their canoes for almost half a mile through a pine forest, now adjoins McLain State Park, where families gather in the evenings for lakeside barbecues.

The Grand Island Chippewa noticed that as they traveled westward the Chippewa who joined them were increasingly warlike; the difference in outlook between the Grand Islanders, who had never been threatened by the Sioux or anybody else, and the western Chippewa, who were in constant danger, became ever more apparent. The divergence extended even to the origins of the tribal name Ojibway. The western tribesmen said that the term came from a combination of the words *ojib*, meaning "puckered or drawn up," and *ub-way*, meaning "to roast." The Ojibway were so fierce, they explained, that when they captured enemies they burned them alive, roasting them until they puckered up. The Grand Islanders were horrified by this explanation and replied that in their tradition the name came from the way in which the tribe made their moccasins, with puckered seams lengthwise. The western Chippewa laughed at this pacific explanation. They were proud of being the most fearsome of all the Chippewa; one group even called itself Mukimduawininewug, Men Who Take by Force, or Pillagers. The Grand Islanders could not help but wonder if their critics had been correct when they implied that the islanders were not really Chippewa, but strangers. Autumn Duck assured them that not all Chippewa agreed with the Pillagers, and that the islanders could retain their own views about war and still be Chippewa. But the islanders did realize that their fears that the western Chippewa might eventually attack them if they refused to be allies in war were not entirely groundless.

Eventually the war party reached the western end of Lake Superior, and there they left their canoes in the care of a Chippewa village and headed west on foot to meet the Sioux, who knew of their approach and were moving to meet them. The exact place where the Chippewa and the Sioux came into contact is unknown, but probably it was somewhere in present-day Minnesota. Some accounts place the battle site close to the Pomme de Terre River, which flows from near the present town of Fergus Falls to the Minnesota River, about a hundred miles to the south. When the first Chippewa scouts sighted the Sioux they returned to the main body with alarming news: while there were dozens of Chippewa, there were hundreds of Sioux. Nonetheless, the Chippewa decided to proceed with the battle. They had come a long way, they were not easily frightened, and they knew that they had more guns than the Sioux.

The leading chief of the Chippewa was Aishkebugekozh, or Flat Mouth. A shrewd, tough, and stern-looking man, he was known as a fierce military leader of the most untamed band of Chippewa, the Pillagers, who lived near Leech Lake. He derived his name from his thin lips extended in a straight line over a prominent jaw. The French called him Guelle Platte, which translates as Flat Muzzle or Flat Snout. Certainly, he was formidable in appearance. Doubting the war abilities of the inexperienced Grand Islanders, Flat Mouth told them to fight a defensive battle against the Sioux and to hold one piece of ground at all costs. This area included a shallow cave or cavern, large enough to hold all thirteen islanders at once. With several entrances and exits, it was more like a very large, mostly covered trench than a cave. It may have been used in previous wars, since it was similar to the "earthen wigwams" often used by the plains Indians for shelter and defense.

In preparing for the battle the Grand Islanders took out their carefully preserved paints from their birchbark containers and unpacked their war dress. They painted one another carefully, recalling all the details they had been taught by the women who first devised their fearsome countenances. The other Chippewa were astounded by their appearance. Several of them, perhaps sensing that the coming battle would result in heavy losses, apologized to the islanders for having called them cowards in past years.

The battle began at daybreak, the Sioux attacking in seemingly unlimited numbers. The Chippewa fought with great spirit and valor, but although they had the initial advantage because of their firearms, the Sioux had learned how to fight the better-armed Chippewa, and they held off from an all-out frontal attack until they had repeatedly drawn the fire of the Chippewa and exhausted their ammunition. Then they swept over their foes. The entire line of Chippewa warriors was driven back except the Grand Islanders, who held fast in the cavern. The Sioux were frightened by their alarming appearance, and it was not easy to dislodge them from their little fortress.

A pause came in the battle. No Chippewa were anywhere to be seen except the Grand Islanders, now surrounded by hundreds of Sioux. One of the Sioux shouted that he was a chief and wished to speak to the Grand Island chief. He was permitted to come into the cavern.

The Sioux chief spoke directly to Autumn Duck, but all the other islanders could hear him. "Your position is hopeless," he said. "You are completely surrounded. You have so few warriors that I can almost count them on the fingers of my hands. I have so many that they are more numerous than the feathers on a flock of geese, and there are even more nearby whom I can call on if I wish. You are brave and fearsome fighters, among the most terrible I have seen, but you have absolutely no chance. We can kill you all very quickly.

"And we have every right to do so," he continued. "You are on our territory; we are not on yours. You are doing what the Chippewa always do, trying to push us farther westward. But now we have our revenge, and we have already taken many Chippewa scalps. All the other Chippewa have fled.

"However, there is no need for you to die. We respect you. We have heard that you did not want to make war against the Sioux. We have even heard about your beautiful island home. Go back to your mothers, sisters, wives, and children. If you surrender and promise never to fight the Sioux again, we will permit you to leave immediately. We urge you to see reason and return safely to the Chippewa lands. The alternative is certain death. We will take no joy in killing you, but if you do not immediately surrender, we will slay every one of you, showing no mercy." Autumn Duck replied that he needed time to discuss it with his companions, and the Sioux chief stepped out of the cavern to give them privacy.

The first islander to speak favored accepting the offer. "We never wanted this war in the first place," he observed. "And the Sioux chief is right; we are on their territory. Yet we do not want their land. We want only to return to our home. Let us do as he says."

The second islander was more dubious: "The Sioux chief speaks well, but why should we believe him? If we put down our arms and come out of the cavern, they will fall upon us and have an easy victory. We should not trust him."

The third islander had yet another view: "It really does not make any difference whether the Sioux chief can be trusted or not. Our fate will be the same whether we fight or surrender. If we fight, we will die. If we surrender, even if the Sioux chief keeps his word, we face disaster. Our surrender will confirm all the suspicions of the other Chippewa that the Grand Islanders are cowards. For years we refused to take part in any war parties. Then when we finally assented and got into our first battle, we immediately gave up! If we return to Grand Island we will be known as the Chippewa who surrender. At the moment we still have no casualties. If we all return safely from a battle in which many lives were lost, our tribesmen will soon learn how we did it. They will turn on us and drive us out of our island home. Our choice is not between life or death, but between death with shame or death without shame. We must fight and die."

The tragic wisdom of the third view impressed them all. The islanders agreed that there was no alternative but to fight, and to fight as bravely as possible. Autumn Duck called the Sioux chief back into the cavern and told him of their decision. The chief said that he was sad to hear their answer. He placed his hand on Autumn Duck's shoulder, said, "Farewell, my brave brother," and left. The islanders readied their weapons for what they knew would be a final, cataclysmic battle. There was no ammunition left for the two guns, so they knew the fight would be a traditional one, both sides with weapons they had used for centuries.

Then a fourth islander spoke up. "I agree with the decision," he said. "But if we are all going to die, our families and descendants should know about it. It should be a part of our history and heritage. We need a witness, someone who will return to Grand Island

and tell our friends, and all the other Chippewa, how bravely we fought and how we died."

This view also made sense, and it was obvious that the only islander who had a chance of escaping and bearing witness was Little Duck. Autumn Duck told his son to try to escape, to run faster than he had ever run before, faster than even he thought he could, and to return to their island with news of the fate of his comrades.

To conceal Little Duck's exit from one end of the cavern, Autumn Duck emerged from the other and emitted a petrifying war cry that astounded everyone—the Sioux, the islanders, and even Autumn Duck himself. The formidable utterance, and the sight of what was the most terror-inspiring warrior any of the Sioux had ever seen, momentarily paralyzed them. Like the mainland Chippewa when they first saw the islanders in war dress, the Sioux felt the ground tremble beneath them. At that moment Little Duck slipped out from the other end of the cavern, hiding behind rocks and trees, and into a dry creek bed. But the Sioux quickly recovered, and dozens of them spied Little Duck. Venting great whoops of alarm, they set after him.

In the creek bed Little Duck had found a place where there was room to run, and run he did. He threw his head back, looked at the sky, pretended that he was on Trout Bay, and ran faster than he ever had before. He felt the hopes and strengths of his friends in the cavern flowing into his muscles, into his entire body. Then, at this ultimate pace, his head commanded, "Faster, faster," and his legs obeyed. The Sioux were wonder-struck at the speed of this Chippewa boy. They loosed clouds of arrows at him, but aiming on the run at a speeding target they failed to hit Little Duck. Hearing the arrows striking the dry mud and rocks around him, he sped even faster.

Little Duck ran for a long time, until he was sure that the Sioux were far behind. Then he stopped, waited, and listened. Hearing no sound, he began to retrace his path toward the cavern but was careful not to overtake his pursuers, who were also returning to the battle site. About an hour later, at nearly midday, Little Duck found a place on a bluff where, hiding behind some bushes, he could see the cavern in the distance but not be detected himself.

The Battle of the Cavern was beginning, with throngs of Sioux gathering on every side. As they attacked, Little Duck caught glimpses of the Grand Island Chippewa furiously fighting, easily identified because of their awesome war paint, and once or twice Little Duck was sure he glimpsed his father.

From the beginning there was never any doubt about the outcome of the battle, but it possessed a quality that terrified the Sioux. In addition to their threatening appearance, the Grand Islanders had an advantage in warfare the Sioux had never before confronted: knowing that their deaths were inevitable and imminent, they fought without consideration of consequence, without heed for their safety. They struggled with a ferocity that gave each of them the strength of the mythic monsters they represented. Originally embarrassed by the phantasmagoric forms their women had invented, later frightened by them, they had now fully incorporated them and were empowered by them. Their war axes traced killing arcs through the air again, again, and again. Dead Sioux at the entrances to the cavern became outer fortifications: those climbing up behind them were impaled by the spears of their own tribesmen, now in the hands of the islanders. All the defenders soon sustained terrible injuries, which seemed only to energize them. The islanders' blood mingled with that of their victims and with the red paint on their faces and arms. They never paused to take a scalp, never whooped when they landed blows, never moaned when they absorbed them, never asked for mercy. Although they fought violently, they displayed no vengeance or hatred toward their foes. It was as if they were in a drama in which they had been assigned the roles of legendary warriors, and they knew exactly how to play their parts. They fully understood that their deaths were a part of the script, but they also knew that the prelude to that finale must be as long and glorious as possible. All of them, and Autumn Duck in particular, hoped that there was an audience. They knew that if that spectator existed, he would preserve their story forever.

The audience was up on the bluff. Little Duck watched the battle through most of the afternoon, but as the sun approached the hills around the cavern, he could see the Sioux walking in and out of the entrances and exits of the fortress, carrying the bloody scalps and weapons of the Grand Island Chippewa. When finally they emerged from the cavern with the scalp of his father at the end of a long pole, Little Duck knew that all the Grand Island warriors were dead.

Little Duck turned back on the trail once again, this time staying in the bushes to avoid detection. The Sioux remained at the battle site, and as he began his long journey to Grand Island, he could hear their victory cries receding in the distance. He found a hiding place until nightfall, and then started eastward.

For several days he traveled only at night; then, as he neared

SIOUX WITH CHIPPEWA SCALP

Chippewa territory, he dared the sun. He lived mostly on berries. He ran and walked, ran and walked, covering territory at a rapid pace. When he neared Lake Superior and the Chippewa village where they had left their canoes, he felt that he was safe; the triumphant Sioux might be advancing into Chippewa territory, but they were almost surely coming at a slower rate.

The women and children in the Chippewa village greeted him and showered him with questions about the other Chippewa warriors, especially those from their village. Little Duck told them that he did not know anything about the others, only that his tribesmen from Grand Island were all dead. The Chippewa gave him strips of dried deer meat and smoked fish and helped him launch a Grand Island canoe into Lake Superior. It was not his father's, which was too big for him to handle by himself, but one of the smaller ones. Even it was too large and heavy to carry alone for any distance.

Little Duck began the long, frightening, and lonely voyage back along the southern shore of Lake Superior. Each night he slept beside his canoe on the shore of the great lake. In the forest he could hear the sounds of many animals. To avoid bears, he would climb a tree before going to bed and suspend his small basket of food high in the branches. Each day he feared that the wind would suddenly rise and wreck his canoe with high waves, or blow him far out in the lake. Once, near the island of Moningwunakauning (now Madeline Island in the Apostle chain), a wind from the south did suddenly increase in force, and Little Duck was almost certain that he would not be able to make it to shore. At the most frightening moment, as he paddled with all his strength, he almost lost hope. Then he remembered the story of Mishosha and his magic canoe. Little Duck slapped the side of his canoe and shouted *chemaun poll* at the top of his voice. The canoe appeared to gather strength, and as it approached the shore, where the water was protected from the wind, the waves subsided. Little Duck reached a pebbly beach, quickly jumped out, and pulled the canoe high out of the waves, trying to avoid damaging it on the rocks. Once safe, he breathed a prayer of thanks to Mishosha and fervently wished to be at home on his enchanted island.

Made cautious by this experience, for the rest of the trip Little Duck stuck closer to the shore, and on days when the wind blew he remained in his camp of the night before. Realizing that he could not get through the Keweenaw portage alone, he took the canoe around the long peninsula. Similarly, at Chequamegon, Keweenaw, and Huron bays he did not cut across open water as his father and friends had done, but followed the shore the long way. These detours, together with the fact that he was traveling more slowly than when two or three people were paddling a single canoe, made his homeward trip much longer than the westward one had been.

When at last Little Duck reached Grand Island he told his dire story to his mother and the other women, the children, and several old men. Grief seized the entire island. The women set up a mournful wail and rubbed their faces black with charcoal. They shrieked and moaned the rest of the day, tore their hair, and sprinkled white ashes on their heads. The old men stuck thorns and needles through their arms and the flesh of their chests. One of them, the ancient Line of Thunder Clouds, sang a song in words so archaic that Little Duck had difficulty understanding them, but he caught the lament, inserted among old undecipherable phrases: "Dear fellow companions, why have you gone so soon to the land of shades? O would that I, aged man, could have gone in your place!"

Everyone knew that the tragedy for the island was even greater than the loss of husbands, brothers, and sons for individual families. It was a collective calamity, since every able-bodied male on the island had been killed except Little Duck, and he was still almost a boy. The future of the Grand Island band of Chippewa was dark. Who would hunt the game, provide the food, protect the village, father the young?

Little Duck's mother, Sound of Wind in the Trees, was overwhelmed by the news of her husband's death, but she managed to show her pride in her son. She told him that after what he had done it was no longer appropriate for him to be called Little Duck, a child's name. He was now a grown man. Henceforth he would be known as Gashkiewisiwin-gijigong, or Powers of the Air, because by running so fast he had proved that he had the powers of the air, the force of the wind. And from that time on, all the remaining Grand Island Chippewa addressed him by his new name and treated him with great respect.

Powers of the Air did not at first notice, but soon after his mother renamed him, she left the mourning crowd and went down to the water's edge. There she took a light canoe, one that she had made herself several years earlier, and paddled across the large bay toward the mainland, but going east, rather than south, where the land was closest. She passed near the south end of the Thumb (where, about seventy years later, the white men would build South Lighthouse) and went toward the Painted Rocks, now gleaming ahead of her in the late afternoon sun.

But Sound of Wind in the Trees did not go all the way to the colored cliffs. She stopped at the first outcropping she encountered, a nameless cliff, now badly eroded, about a mile west of what was later called Miners' Castle. She knew that the cliff was too steep to climb directly up to the outcropping that was her goal, but several hundred yards west of it a small stream had cut a ravine into the rock that afforded a path upward. A little bit west of the ravine was a cleft in the rock at the water's edge, a curious indentation in the rock face that can still be seen. It was just wide enough for her canoe to slip in. There she left it, then climbed up the low cliff above by grasping the branches of a tree that had fallen toward the water. Grasping bushes to aid her ascent, she crossed back and forth ever upward along the ravine, and at last stood on top of the cliff. The cliff slanted gently back from the shoreline until, about two-thirds of the way up, it suddenly projected outward, like the prow of a ship. The spot where she now stood was over both air and water. To the north, she could see the entrance to Trout Bay and the eastern side of Grand Island. To the west she saw a short peninsula, now called Sand Point, with low pine trees stretching into the channel between the mainland and the island. This spot, the place where the mainland Chippewa buried their dead, is today the headquarters of the Pictured Rocks National Lakeshore, housed in an old building of the Life-Saving Service. To the east rose the stark con-figuration of Miners' Castle, a spot favored much later by tourists wishing to catch a glimpse of the Pictured Rocks from land.

The spot where Sound of the Wind in the Trees stood had a place in old Chippewa tales. According to some, it was the place of encampment of the niece of Manabozho, a mythic figure whom the whites would call Hiawatha. But for Sound of Wind in the Trees, the location had a deep personal meaning. When she had been a young woman approaching the age of marriage she had come here, as was the local Chippewa custom, to seclude herself for twenty days. She knew that when she returned to the village, the first eligible young man to see her would turn weak in the knees and fall inescapably in love. When she returned she had made sure that Autumn Duck was the first man to see her.

She looked down at the water, far below, and thought about those times. She had come here without a specific purpose, except, now that Autumn Duck was dead, to return to the spot where her love for him had blossomed. As she stood on the narrow promon-tory and gazed in her grief toward the painted cliffs to the east, a long stretch of the most legendary sites on the shore of the lake, something happened. Did she become dizzy, slip, and fall? Or did she throw herself into the water below? No one would ever know.

Her body, lodged in the rocks at the water's edge, and the canoe, drawn up in the rock cleft, were not found until the next day.

A SONG FOR THE DEAD

POWERS OF THE AIR knew that he had a responsibility to preserve and tell the story of the fates of the Courageous Twelve. He went for help to Nahbenayash, Line of Thunder Clouds, who knew more old legends about the tribe than anyone else on the island and who also was expert in composing sequences of words that fitted the repetitive cadence of Chippewa songs. Thereafter, Powers of the Air and Line of Thunder Clouds met many times at the south point of the island. They sat together on a rock overlooking the strait separating the island from the mainland and composed an epic song filled with praise and sorrow for the Grand Island men who had voluntarily sacrificed themselves in a war they did not believe in to preserve the honor of their band. The song told how the Grand Island Chippewa had always been peace loving but, when forced to join their tribesmen in a campaign against the Sioux, had refused the offer of their lives from their attackers. They had fought more valiantly and had died with greater understanding of their lives and deaths than any Chippewa in memory.

Line of Thunder Clouds told Powers of the Air that the story was unusual among Chippewa songs because it was not a celebration of an individual's heroic feats or a tale involving dreams, love, or games, but a lament about a group. It portrayed a unique view of the world—the Grand Islanders' way of life and their unwillingness to accept the mainlanders' views about warfare. Despite its bloody ending, it seemed strangely peaceful, in harmony with how the Grand Island Chippewa had always lived. War was described as a harmful and mistaken activity. The Sioux, the traditional tribal enemies usually portrayed in Chippewa tales as fierce beasts, were depicted as people who offered the Grand Island men their lives if they would only leave Sioux territory. When the Grand Islanders refused, the Sioux attacked not vengefully, but as brethren fighting for the same reason as the Grand Islanders. They were caught in a situation that gave them no choice, and when required to fight, they were as relentless in offense as the Grand Islanders were in defense. There were no individual heroes on either side. Although Powers of the Air had been named a "brave warrior" *(ogitchida)* by his family and friends, Line of Thunder Clouds could recall no other instance in which a Chippewa warrior had won this title not for valor in battle but for success in running away from one. The old storyteller was doubtful that the song they were composing would be enthusiastically received either by the Chippewa on the mainland or by the Sioux. The epic's survival rested with the Grand Islanders, the only people who understood its message.

After Line of Thunder Clouds and Powers of the Air finished their lengthy song, Powers of the Air began to sing it to the other islanders, at first hesitantly, then with increasing assurance. The women learned the cadence quickly, joining in with a low chant and beating out the rhythm on the birchbark walls of the lodges. Later they made instruments for accompaniment. Charring and scraping the inside of a basswood log about the length of a forearm, they fashioned a drum, fitting a wooden disk to one end and a piece of deerskin to the other. In one side they drilled a hole into which they poured water from a birchbark container. The water imparted a peculiar resonance, a sound that was soft and low yet audible in all the lodges. One woman made a flute *(pibbegwon)* from two hollow semicylindrical pieces of cedar by joining the wood with fish glue and drilling eight holes. She then drew a wet snakeskin over the instrument; when dry, it held the pieces tightly together. Other women made rattles from gourds and pebbles, which they shook rhythmically as they hummed and chanted.

Powers of the Air sang the song many times, and he noticed that each time he did so he almost involuntarily changed it. It was like a living thing that demanded revision and embellishment. The chorus of women standing to each side of him encouraged him to describe the expedition and the battle in ever greater detail, expertly adjusting their accompaniment to each variation. As time went on and the modifications continued, Powers of the Air and all the others listening and joining in were no longer certain about what had been in the original account and what they had added, but it no longer mattered. The story had become a legend, and what it meant was more important than what had occurred.

The Chippewa women grieved that they could not give their departed husbands, sons, brothers, and fathers appropriate burials. According to tribal custom, carved markers indicating the clan of the deceased were often placed at individual graves. Since no such

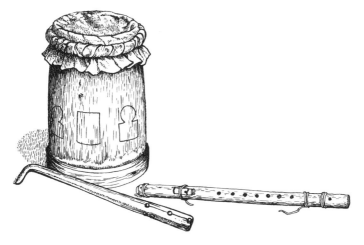

WATER DRUM AND FLUTES

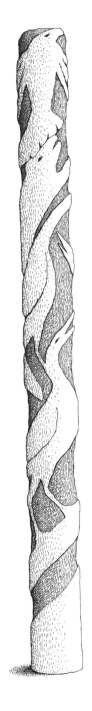

TOTEMS OF DEAD
GRAND ISLAND WARRIORS

graves existed for the Twelve, the women decided to make a collective marker commemorating the legend. Into a cedar pole about ten feet long, they carved in partial relief in a spiral the *do-daim*, or totem, of each man who had died: a crane, a loon, a bear, and most of the other totems of the Grand Islanders. At the bottom of the pole was a small duck, and at the top a soaring eagle, representing the transformation of Little Duck into Powers of the Air. They painted the background red and left the figures themselves in the natural wood. The result was a striking monument, not at all like the totem poles of Native Americans on the western coast of North America, but instead a large-scale version of the traditional Chippewa grave markers. When the women had completed the marker they placed it upright in the burial ground between Eagle Lake and the great lake, facing to the west to encourage the spirits along their way. For four days they kept a fire burning nearby.

A PARTRIDGE FOR PEARL-WHISPER

FOR THE FIRST year or two after Powers of the Air returned from the Battle of the Cavern, the only bountiful aspect of his existence was his romantic life. With the exception of some small boys and old men, he was the only male around. Strikingly handsome and the bearer of the great title *ogitchida,* he was adored by both the young girls and the older *anishinabekwe*. As all the Grand Islanders struggled to find enough food to survive, the flirtations between Powers of the Air and the younger women were often the only diversion. Soon everyone's attention focused on who would become his wife. Powers of the Air seemed in no hurry, although he was frequently seen disappearing into the woods with one or another of the young women.

The choice was limited. Only seven or eight women were close to his age, and of these, three or four had the same totem, the bear, as he did. According to Chippewa custom it was as unthinkable to marry one of them as it would be today to marry one's sister.

Among the other four or five who were eligible, Powers of the Air seemed to have little preference. He liked them all, and they all liked him.

All the eligible young women wore about their necks little leather bags with love charms consisting of two small wooden figures representing a man and a woman tied together. To be effective the charms also had to include some hair from the head of the desired man. The young women would try to sneak up on Powers of the Air while he was asleep to grab a few strands, but he was extremely alert, and laughingly escaped. After a while all the bags contained hair from his head, even though no one had seen the young women getting it. The older women knew that the hair had been obtained on those amorous occasions when the young women had disappeared with Powers of the Air into the woods.

The senior women of the tribe were lenient for a year or so. They, too, were without men and they understood. But eventually they began to complain that it was time for the young man to settle down with a wife. Powers of the Air replied to their affectionate criticism by saying that he would marry the first woman who could catch him while he was running. This offer caused great merriment, since everyone knew that no Chippewa on the south shore of Lake Superior, man or woman, could catch Powers of the Air in a footrace.

Nonetheless several of the young women tried. Powers of the Air would run slowly until his pursuer was almost in reach, then spurt ahead. But he did not take the race seriously, and sometimes he would turn and chase the women. Since the women understood that he was just playing, and that it made no sense for him to catch them, they would also turn and dash away, laughing. Powers of the Air would pretend that he could not catch them.

Pearl-Whisper, a wise and beautiful young woman, decided that she would interpret Powers of the Air's promise in her own way. She carefully studied his habits and saw that every morning he ran the length of Trout Bay beach several times, always doubling back at the east end where the sand met several large rocks. She decided to

take her daily morning bath behind those rocks, carefully conceal-
ing herself from his gaze but splashing the water noisily enough
that he would hear her.

After a few days Powers of the Air began making his turn at the
east end more slowly, running a bit higher up the little hill beyond
the beach to gain a better view over the rocks. One morning he ran
to the east end but did not return. He disappeared until nearly midday, when he and Pearl-Whisper were seen walking back along the
beach, hand in hand, their feet splashing in the water. That afternoon Powers of the Air took his bow and arrow and went out hunting. He could not find the deer that he wanted, but time had suddenly become scarce for him, so he returned with a plump
partridge. He walked to the lodge of Pearl-Whisper's mother and
presented her with the partridge. She asked him to stay and share
the bird with them. From this sequence everyone knew that Powers
of the Air had proposed, Pearl-Whisper had accepted, and her
mother had approved. From then on, Powers of the Air and Pearl-
Whisper lived together.

The marriage was a momentous event in this precarious time for
the Grand Island Chippewa, and they decided to celebrate it in the
most extravagant way possible, with a Stone Pot Feast. This was not
an easy undertaking, since food had become scarce. Traditionally
the fish and beaver tail boiled in the stone pots were not smoked or
dried, but fresh, meaning that the fish had to be caught and the
beaver killed the previous day. Obtaining such food in sufficient
quantities had not been a problem when there were a dozen or
more men to hunt and fish while the women gathered the vegeta-
bles, herbs, and mushrooms. But now there was only one man, and
as one of the two most honored guests of the feast it did not seem
right to push him to exhaustion. To make matters worse, the Grand
Islanders had lost all but two of their canoes in the war party against
the Sioux, and there had been no time to build others. Fishing
would not be easy. Also, canoes had always been used to go up the
coast to the northern end of the island where the stone pots were located. The two canoes would not hold the forty or so Chippewa remaining in the village.

Fishing, even with only two canoes, was nevertheless easier than hunting beaver. The Chippewa decided to forget about the beaver tail and concentrate on the fish. Powers of the Air and three women went out in the two canoes. The other women and the children scoured the woods for berries, herbs, and mushrooms. The feast was being held in early October, Falling of the Leaves Moon, and the gardens still contained a little corn and potatoes.

When the time came to go to the feast site, several small children and old people were carried in the canoes. All the rest of the villagers hiked up the ten-mile trail along the center of the island, carrying the supplies on their backs. The forest was illumined with the orange and red of the maple and beech leaves, and the colors lightened the hearts of people who had suffered much and who looked forward to a cele-bration, however brief.

Several of the older women sensed, but did not say, that this would probably be the last Stone Pot Feast ever held on Grand Island.

LOVE CHARM

They knew that they had not stored nearly enough food for the coming winter, and that a crisis was coming. The future of the Grand Island Chippewa was in jeopardy. But these fears were put aside for the Stone Pot Feast celebrating the marriage of Powers of the Air and Pearl-Whisper.

The islanders gathered on one of the stone shelves at the base of a cliff where there were several pots suitable for boiling the food. One of the pots still contained heating stones from the last feast held several years earlier, before their men had been lost at the Bat-tle of the Cavern. The women cleaned out the pots and filled them with water while Powers of the Air started a fire, using a fire-drill made of a small bow that he sawed back and forth, twirling a drill stick in a hole in a piece of wood. He used crumbled dry leaves as tinder, and within a few minutes he had a small flame, which he fed with tiny pieces of driftwood. Some of the children brought more wood, and the flames grew rapidly into a bonfire that heated the stones.

After the feast of boiled whitefish, trout, and mushrooms the islanders gathered around the fire to sing songs and tell stories, as was their custom at the end of their Stone Pot festivities. All the Chippewa knew by heart the most familiar songs and stories, and once the ritual gained momentum and began to infect the participants emotionally, they would sing and talk simultaneously, in four or five groups, as if each was paying no attention to the others. One person would stand up and begin to speak, while a few feet away another would do the same, telling an entirely different tale. The first person might finish and sit down, whereupon the other members of that group would start to sing a song even though several persons in other groups were still standing and speaking. To the outside observer it appeared to be an orderless cacophony, but the Grand Island Chippewa all knew the various stories and songs so well that they needed to catch only a few words from any one to know exactly what was going on there. By doing all this simultaneously, they could run through their entire repertoire, renewing their traditions, in little more than an hour.

The song of the Battle of the Cavern was still so new, and the sentiments it evoked so fresh, that everyone wanted to hear every word. After the familiar confusion had ended, they all gathered around Powers of the Air to listen to the song again, with Pearl-Whisper playing the flute and the other young women who had lost out in the love competition providing accompaniment with the water-drum and gourd rattles. The mournful but heroic song united the emotions of the entire group, and when it was finished the islanders felt better prepared for the hard times that they knew were coming in the cold months ahead.

HUNGER

THE WINTER THAT followed, the second after the Battle of the Cavern, was the worst in the history of the Grand Island Chippewa. By late December, Little Spirit Moon, the food was exhausted except for some dried blueberries. By the end of January, Great Spirit Moon, the hunger began. The old people refused to take food from the children. Line of Thunder Clouds was the first to die. But several children soon succumbed to cold and sickness.

Each day Powers of the Air went out hunting on snowshoes, carrying his spear and his bow and arrows. When the Chippewa had pursued deer in groups of ten or twelve hunters, they had been able to surround the animals and herd them back toward the lodges, "running down the game" in the traditional way. But even in the heavy snow, the deer could avoid Powers of the Air hunting alone. He was forced to shoot his arrows from afar, and they missed.

One evening after Powers of the Air again returned empty-handed, several young women came to his lodge and asked to speak privately with Pearl-Whisper. After winning her over to their plan, they presented it to Powers of the Air. They were good runners, as they had proved when he had chased them. Why could they not

YOUNG POWERS OF THE AIR

help him on the snowshoe hunts? Powers of the Air reminded them of what they already knew: according to Chippewa custom, women could gather berries, carry firewood, cook the meals, skin animals and tan the skins, build canoes, and tend the gardens, but they were never allowed to hunt. Hunting was reserved for the *in-inig*, the males. The women replied that their small group was now in danger of starvation. Besides, the Grand Islanders had never been like the mainland Chippewa; they had never engaged in war even though the mainland Chippewa waged war unceasingly against the Sioux. It was now clear that the one time when the islanders had deviated from their peaceful ways and joined their tribesmen on a war party, they had made a terrible mistake. Perhaps they should return to doing things their own way. They would form a hunting party made up of women and one man. Powers of the Air listened and said that he would think about it overnight.

The next day Powers of the Air agreed that the plan was worth trying. Six young women, those who were able to run for a long time on snowshoes, including Pearl-Whisper, prepared for the hunt with Powers of the Air. Unaccustomed to using bows and arrows, they carried short spears and knives. Powers of the Air had a bow and arrows and a long spear.

Lacking experience and energy, the hunting party still faced great difficulties. Powers of the Air, a very young man, was not nearly as good a hunter as his father or the older men had been. And not having eaten anything for days except dried berries and a few roots, everyone was weak. But the cheerful and laughing young women raised morale. As the group floundered in the snow, trying to herd the deer, some of the old flirtatiousness between Powers of the Air and the women returned as they all shouted and joked among themselves. Pearl-Whisper did not mind; these were extreme times in which anything that raised the spirit was precious.

Time and again the group would surround a deer, shouting to frighten it and trying to drive it where they wanted. Again and again the deer would elude them, running around the line they had formed or leaping through it as they ineffectively threw their short spears. Even with seven hunters, the group was too small for the task. Each evening they would return weary and empty-handed to the lodges.

Each day, though, they were becoming more skilled; and finally, in mid-February, Sucker Moon, they ran a deer to exhaustion not too far from the village. For a moment hunters and hunted alike lay panting in the snow, their sides heaving from the exertion. Then Miskcoowahgegoneance, Little Red Flower, discarded her snow-shoes and crawled through the snow to the deer while the animal, its eyes protruding with fear, stared transfixedly at her. At last Little Red Flower could feel the deer's heavy breath on her face. Had it been a buck rather than a doe, it could have impaled her. She threw an arm around the deer's neck, drew her knife, and slit its throat. A small crimson stream spouted onto the snow. Little Red Flower cupped her hands under the cut and drank the hot liquid, besmearing her face, hands, and deerskin shirt. The hunting party cheered. Hearing the noise, several older women came from the lodges to drag the deer home.

As the women approached, Powers of the Air remembered what his father had taught him to do at such moments. Leaning heavily into the snow, he murmured so softly that only the hunters closest to him could hear his words: "*Wawashkeshi, ninimoshi.* Deer, my cousin, forgive us for what we have done, and thank you for providing us with sustenance." The exhausted hunters dragged themselves home. By the time they reached the lodges the older women were already dressing the deer.

That deer saved the lives of the Grand Island Chippewa. They consumed every bit of it, cracking its bones and eating the marrow. They were proud that the islanders had returned to doing things their own way, and Little Red Flower was celebrated as a great hunter.

Soon the deer was gone, and food was again scarce. Now it was early March, Breaking Up of Snow Moon, and the islanders knew that if they were to escape to the mainland where other Chippewa might help them, they must go now. In a few days the ice between the island and the mainland would soften, but many more weeks would pass before the trip could be made by canoe. It was their last chance to walk out. They gathered a few possessions, wrapped themselves in beaver robes, and began the trek across the ice.

The arrival of about thirty additional mouths to feed was not a welcome event for the mainland Chippewa. Spring was always the time of greatest scarcity for the lodge dwellers around Lake Superior, even with sufficient men for hunting. Stored food from the previous year was exhausted, the diminishing snow cover and numerous wet spots made running down the deer on snowshoes more difficult, and the soft and dangerous ice on the lakes prevented much fishing. But the mainlanders knew what had happened to the men of Grand Island, and they sympathized with the survivors. They even felt a bit guilty, since for years they had ridiculed the island men and had helped to coerce them to go to their deaths. The mainlanders let the islanders enter their lodges and shared their food with them.

As time went on the Grand Islanders were absorbed into the mainland population. Many of the young women married mainlanders. When summer came the islanders returned occasionally to their home to gather berries, but in the evenings they usually came back to their mainland lodges.

FROM CULTURE
TO COMMODITIES

A FEW SUMMERS after the islanders left their home, something happened that hastened the absorption and disappearance of a separate Grand Island Chippewa culture. A small group of white men from the American Fur Company arrived in the bay and established a fur trading post on the south end of Grand Island. The fur trade had existed on the south shore of Lake Superior for many years, but earlier posts had been scattered along the mainland, and the relative isolation of the islanders had protected them from dependency on the trade. The new arrivals were interested in Grand Island itself, especially the beaver colony at Echo Lake.

The white fur traders brought many metal traps and offered their use to the Chippewa, saying that they could pay for the traps with pelts and that afterward they could trade pelts for many other valuable things—guns, whiskey, cooking utensils, mirrors, trinkets, knives, shirts, trousers, and tools.

Many Chippewa accepted this proposal, coming to the island

from as far away as the northern shore of Lake Michigan. Among the mainland Chippewa who visited the island regularly the chiefs included Sabboo and Oshawonepenais (South Bird) and, later, Omonomonee and Kishkitawage. From this time on, Grand Island became an integral part of the mainland Chippewa culture. Some of the original islanders, including Powers of the Air, came to the island to help in the beaver hunts, but their separate identity was soon lost. By this time many of the original Grand Island women had married either mainlanders or white men who were moving into the area.

The slaughter of the Echo Lake beaver proceeded rapidly. Chippewa women came over to the island with the trappers to dress and trim the hides and stretch them in great circles on wooden racks. So many pelts were collected that the white traders built a wooden crane at the southernmost point of the island, where the heavy packs of furs were loaded into large canoes for the trip to Sault Ste.

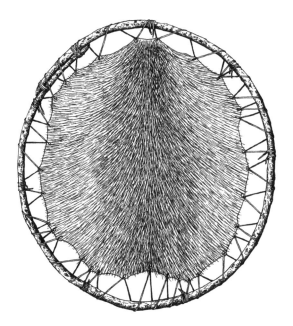

BEAVER PELT BEING STRETCHED

Marie. The crane was located next to the rock where Powers of the Air and Line of Thunder Clouds had spent many hours composing their song.

The Chippewa soon became dependent on the trading post for clothing, liquor, tools, utensils, cloth, and guns and ammunition. The white fur traders encouraged this dependence; they traded liquor for food in the fall, and then traded the food back to the Chippewa in the late winter and early spring, taking furs in return. The Chippewa were thus constantly driven to increase the harvest of furs, despite the rapidly decreasing population of beaver.

Powers of the Air and Pearl-Whisper were the only Chippewa who made an effort to preserve the separate life-style of the Grand Island band. During the summers they continued to live alone on Trout Bay, where their childhood homes had been. Aware that the old tradition of the Grand Island Chippewa had been one of aloofness from the mainlanders, they continued to live apart to the degree that they could, surviving on fish from Trout Bay and Eagle Lake. The other islanders built lodges next to the trading post, where it was easy to trade furs and fish for the white men's commodities. Liquor was one of the most sought-after items, and it soon had a destructive effect on the Chippewa.

After a few years the beaver population on the island had been devastated, and the American Fur Company moved farther west,

FUR TRADING CABIN WITH LIFTING CRANE

abandoning their cabins on Grand Island for the use of anybody who wished to occupy them. With the departure of the fur traders, only a few Chippewa continued to come there for the summers. The island was cooler then than their homes on Little Bay de Noc to the south, and fish and blueberries were still plentiful. But with the beaver, their main source of food and warmth, largely gone, it would have been difficult to remain on the island through the long winters.

SPREADING
THE LEGEND

IN THE SUMMER of 1820 a large party of whites set out from Detroit to explore the southern shore of Lake Superior and to seek the source of the Mississippi River. An important political goal of the expedition was to show the American flag in these regions and to counteract British influence, which was still strong around Sault Ste. Marie. Leading the expedition was Lewis Cass, governor of the Michigan Territory, who would later serve as secretary of war and as ambassador to France and in 1848 would be the Democratic Party's candidate for president. In 1820 the young and ambitious Cass was eager to learn more about the territory he governed, which encompassed not only the lower and upper peninsulas of present-day Michigan but also most of the western shore of Lake Michigan, today's Wisconsin. Fewer than 10,000 whites inhabited this large region, and relatively little was known about it. Cass wanted to collect information on its topography, geology, resources, and people. Intrigued by the history and culture of the native people, he took along several interpreters. Cass was also well aware that publicity about his exploits would enhance his political prospects. For that reason he invited a lawyer-journalist to join the party, James Duane Doty. Still others accompanying him would later be influential in determining American policy in the Northwest Territory: Henry Schoolcraft, a mineralogist who would become Indian agent at Sault Ste. Marie and an early scholar of Native American culture; David B. Douglass, a professor of mathematics at West Point and later a well-known civil engineer; and Charles C. Trowbridge, a young businessman who would become a civic leader in Detroit. In all, the party included forty men traveling in large canoes built specially for the trip.

On June 15, 1820, the expedition arrived at Sault Ste. Marie, where Cass asked the local Chippewa to cede land on the south side of St. Mary's River, directly across from British holdings, so that he could build an American fort there. The Chippewa refused outright, not only because their sympathies lay more with the British than with the Americans, but also because the land Cass wanted contained one of their cemeteries. Cass made it clear that he was not in a mood to compromise, and an ugly situation developed. The two Chippewa leading the opposition to Cass were the local Boweting chief, Singabawossin (Spirit Stone), and the famous war chief Shingwauk (Little Pine), who had left Grand Island years earlier because of his disagreement with the islanders' antiwar policy. Dressed partially in British army uniforms, Singabawossin, Shingwauk, and an even more militant young chief named Sassaba refused the gifts the Americans offered, thrust a war lance into the ground, stalked out of the meeting, and returned to the Chippewa village, where they raised the British flag. Sassaba urged his tribesmen to attack the Americans. Both sides prepared for combat. The Chippewa, with fifty or so warriors, had the advantage in numbers.

A Chippewa woman married to the most prominent local white trader stepped in and prevented what seemed to be an inevitable battle. Ozhawguscodaywaquay (Green Meadow Woman), wife of John Johnston, persuaded the three chiefs that she possessed magical powers, given to her during an out-of-body vision or trance, that enabled her to avoid calamities. She persuaded Singabawossin, whom she knew well, not to attack the Americans, and Singabawossin managed to pacify Shingwauk and Sassaba. Thus warfare was avoided.

The confrontation reflected a common pattern in the Sault Ste. Marie–Mackinaw region. Quite a few white men who exerted influence in politics, trade, or religion married native women who played important roles in maintaining good relations between the whites and the local tribes. They included John Johnston, John Haliday, William Aitkin, Henry Schoolcraft, James Schoolcraft, Rev. William MacMurry, Benjamin Pierce (brother of President Franklin Pierce), and Edward Biddle (brother of Nicholas Biddle, president of the United States Bank). These marriages promoted significant strategic considerations, enabling the white men to get what they wanted without having to wage war.

After consolidating their hold on Sault Ste. Marie, the American expedition headed westward along the south shore of Lake Superior. On June 21, 1820, they arrived at Grand Island, where, in contrast to their experience at Sault Ste. Marie, they met with a friendly reception from the Chippewa. Cass learned that the American Fur Company had left a year or two earlier, and that there were now no whites in the area except for the occasional explorer or hunter. The Chippewa also told him that they moved back and forth between the mainland and the island according to the season, but that not many years earlier there had been a permanent Chippewa village on the island. When Cass asked why the original Chippewa had left, he was told that he should talk to a young man named Powers of the Air, who lived with his wife and child in a summer lodge on Trout Bay. Powers of the Air, the Chippewa all said, knew some interesting stories about the island. Cass sent one of his men, a French trapper from Sault Ste. Marie who spoke passable Chippewa, to offer Powers of the Air a gift if he would come to the governor's camp and tell him about the history of the island. The trapper paddled his canoe to a spot on the island near the old Chippewa burial ground and then walked over to Trout Bay, where he found Powers of the Air, Pearl-Whisper, and their child living alone. They re-

turned with the trapper to the governor's camp next to the fur traders' cabins.

Cass introduced himself to Powers of the Air, said that he had
heard of some stories about Grand Island, and asked if Powers of
the Air would be willing to tell them to his party. Powers of the Air
replied that the story he had to tell was a long one, that it required
some preparation, and that he would not be able to tell it before
evening. Besides, he continued, it was not a story, but a song, and
he needed help from his friends to sing it in the manner that the
Chippewa would approve of. He asked if he might borrow a canoe
to go after the friends, since his own was over on Trout Bay. Cass
willingly lent him a canoe.

With the canoe Powers of the Air went looking on the mainland
for several of the women who knew how to accompany him and
also for the old basswood drum. Many Grand Island women had
moved away, either to Lake Michigan with the Chippewa or westward along the lake with trappers they had married. But he found
Little Red Flower, now married to a mainland Chippewa, and she
had the basswood drum. He returned with her to the island to get
Pearl-Whisper, who still had the cedar flute.

That evening all the Chippewa on the island built a bonfire near
the lakeshore, where today is a long wooden dock used by the island's summer residents. In June it does not grow dark there until
late. When darkness had covered the shore but the bay still glim-
mered from the glow of the sky, the Chippewa lighted the fire, and
Powers of the Air stood nearby, flanked on one side by Pearl-
Whisper, with the flute, and on the other by Little Red Flower, with
the basswood drum, which she had properly filled with water. Several other Chippewa had gourd rattles. Powers of the Air was
dressed in deerskin. His head was devoid of decoration except for
two eagle feathers that hung downward from the top and back,
emerging beyond his right cheek. The white men gathered around
to listen. Governor Cass sat on a log near an even younger white
man, and between them sat a Chippewa member of their party
who quietly interpreted the words of Powers of the Air.

Powers of the Air realized that these men were much more interested in his song than any of the trappers from the American Fur Company had been or, for that matter, any of the mainland Chippewa. With the gradual dispersion of the Grand Island Chippewa he had been losing both his accompaniment and appreciative listeners. Line of Thunder Clouds had predicted such an end for their song. Powers of the Air decided that the evening was one worthy of his best efforts. He mentally transported himself back to the times about which he sang, the peaceful days of the Grand Island Chippewa before the war party, the trips to the colored cliffs to get the war paint, and the expedition along the shore of the great lake.

When he came to the Battle of the Cavern itself, Powers of the Air had become so much a part of the song that he felt he was there once again watching, and he added details that now only he could recall. He sang so beautifully that all members of the listening crowd, even the white men without interpreters who understood not a word, were rooted in their places. Pearl-Whisper and Little Red Flower, chanting softly and deftly maintaining the rhythm, also knew that this was a special occasion, and tears glistened on their cheeks. They were a part of the song, too, witnessing the end of their Grand Island way of life.

Powers of the Air noticed that the young white man sitting near Governor Cass had his right ear close to the interpreter and, as the interpreter rendered the words of the song into English, was writing on a tablet. When Powers of the Air had finished the song, the group was hushed, for a while, and then the young white man came forward with the Chippewa interpreter. He introduced himself as Henry Schoolcraft; he was a geologist but was also collecting information on the American Indian. Would Powers of the Air talk to him about his tribe?

In the conversation that followed Schoolcraft complimented Powers of the Air on the remarkable song, particularly the rhythm and the verse. But the story itself, intriguing as it was, concerned recent history; and Schoolcraft seemed more interested in the ancient legends of the Chippewa, the sort of thing that Line of Thunder Clouds had known so well. He asked a lot of questions about the mythic Manabozho. Powers of the Air knew many of the traditional stories about Manabozho, and he did his best to tell what he knew, regretting that he had forgotten so much that Line of Thunder Clouds had told him. When Schoolcraft asked about legends that were directly related to Grand Island, however, Powers of the Air was on firmer ground. He knew several of these stories by heart, including that of Mishosha, the magician of the lake. He told Schoolcraft about Mishosha and his magic canoe in great detail, including something that in earlier years he would never have revealed. But now that the original band of Grand Island Chippewa had disappeared forever, preserving its secrets was no longer necessary. To Schoolcraft, then, Powers of the Air unveiled the magic command, *chemaun poll*, which would cause a canoe to speed through the water faster than the wind. Schoolcraft scrupulously wrote the words in his notebook.

Schoolcraft asked Powers of the Air to tell him other stories about Grand Island and the surrounding area, and the young Chippewa obliged with two legends.

The first was about Wawabezowin, an enormous rope swing on a high cliff on the Painted Rocks. Exhilarated by soaring in immense arcs out over the lake, people willingly risked the dangers of the swing. Once an evil hag tried to get rid of her daughter-in-law by encouraging her to view the cliffs from the outermost point in its arc. When the young woman reached this spot and turned in the swing to view the cliffs behind her, she saw her mother-in-law cut the ropes. The daughter-in-law soared far out over the lake, then plunged into its depths. The old woman was sure that this was the end of her and that now she could enjoy all the attention of the family. But the young wife escaped the death planned for her. Sinking to the bottom of Lake Superior, she found the luxurious lodge of one of its ruling spirits, the water-tiger, who fell in love with her and asked her to be his wife. She replied that first she must go back to the Painted Rocks and retrieve her child. Leaving his domain, the besotted water-tiger accompanied her. The young woman's husband killed the water-tiger with his spear, and the couple was reunited. When the mother-in-law saw that her murderous plot had failed, she fled into the forests, never to return.

The second legend that Powers of the Air told Schoolcraft was that of Kaubina and the charmed arrow. Many years ago there was an Ottawa chief named Sagimau who lived on Lake Huron. Sagimau was a great warrior who drove off one of the earlier tribes that had lived in the area and established himself as the undisputed ruler. As time went on Sagimau came to think of himself as the most eminent chief of all the Great Lakes. He could not stand the thought that there might be somewhere on their shores a rival to his fame and power. Yet Sagimau began to hear more and more about a chief named Kaubina, who lived on Grand Island in Lake Superior. Kaubina was not known as a great warrior, but he had enormous spiritual power and was rumored to be a *manito*, or religious leader. He had the assistance of a witch who lived under Lake Superior and referred to him as her grandson. The witch always reported to Kaubina the plots of others against him, so that Kaubina was able to defeat his enemies, usually by outsmarting them.

Sagimau grew more and more jealous of Kaubina and was determined to best him in a contest. Kaubina had a young and beautiful wife whom Sagimau decided to kidnap. Normally the witch would have warned Kaubina of the plot, but Kaubina had offended her by having been in a bad humor and treating her ungraciously when she had recently visited him. Thus Kaubina was ignorant of Sagimau's approach. Using magical powers, Sagimau and his men crossed to Grand Island from the mainland by walking on the bottom of the strait. Deep in the water they met two monster-spirits, whom they appeased with gifts of tobacco. The monster-spirits helped Sagimau and his men reach the island, where they assumed the shapes of large white rocks and pieces of driftwood. When Kaubina's wife went to the shoreline to fetch water, Sagimau seized her and carried her away.

Distraught by the disappearance of his wife, Kaubina begged the witch to forgive him for his earlier inattention and to help him retrieve his wife. The witch took pity on him and used her magical powers to find out that Sagimau was still a day's journey from Grand Island, near the great sand dunes beyond the Painted Rocks. Kaubina immediately went there with his men and demanded that

Sagimau return his wife. Sagimau realized that because Kaubina had found him so quickly he must possess magical powers and that he himself must give back Kaubina's wife. When Kaubina asked Sagimau why he had taken his wife, Sagimau replied, "To see how great a *manito* you were. Here she is; take her. Now that I know your magical powers, we will live in peace."

But Sagimau had not given up hope of besting Kaubina. He knew that to do so he must acquire magical powers greater than those of Kaubina. He went home to Lake Huron and thought long and hard about how to do this. It was known among the warriors of the Great Lakes that there existed a charmed arrow that could strike any target and pierce any tough skin, even that of the greatest beasts. The person who possessed this arrow would rule over all others. This charmed arrow, however, was in the possession of Waubwenonga, the king of the birds, a great vulture-eagle. Sagimau decided that he must get this arrow to defeat Kaubina. Because the vulture-eagle fed on carrion, Sagimau disguised himself as a dead moose in order to attract Waubwenonga. Soon the vulture-eagle appeared and penetrated the moose's hide with the charmed arrow. Thereupon Sagimau transformed himself back into a warrior-chief, seized the arrow, and made off with it.

Sagimau then launched an attack against Kaubina. A titanic battle ensued on the mainland and the island, with each of the two great chiefs using the magical powers that he possessed. For a long time the outcome remained uncertain. The witch under Lake Superior, fearing for the survival of her grandson, gave him a fighting skin, made from the fur of a grizzly bear, that was invulnerable even to the charmed arrow. After a battle lasting all day Kaubina defeated Sagimau. During the combat Waubwenonga regained the charmed arrow and kept it from that time on. In later years the Grand Island Chippewa continued to believe that the charmed arrow was still on the island, kept there by one of the great eagles.

While Powers of the Air was telling these legends to Schoolcraft, others in Cass's party were still discussing the song about the Battle of the Cavern. They demanded that the Chippewa interpreter give it again in English loudly enough that all of them could hear. These conversations went on for several hours, and it was deep into the warm night before Powers of the Air and Pearl-Whisper could lie down and sleep, their child between them. The next morning they walked back to their lodge on Trout Bay, carrying a new ax given them by Governor Cass.

Both Schoolcraft and Cass recorded the occasion of the singing of the song by Powers of the Air. In his journal Schoolcraft described the young troubadour as "a tall and beautiful youth, with a manly countenance, expressive eyes, and formed with the most perfect symmetry—and among all the tribes of Indians whom I have visited, I never felt, for any individual, such a mingled feeling of interest and admiration." The Chippewa epic so impressed Cass that he wrote about it over six months later to John Calhoun, at that time U.S. secretary of war. Referring to the Battle of the Cavern, Cass told Calhoun that in the wars between the Chippewa and the Sioux "instances of courage and self-devotion have occurred, within a few years, which would not have disgraced the pages of Grecian or of Roman history." Yet another member of the party, the journalist James Doty, wrote about the deaths of the Grand Island warriors in an article printed in the Detroit *Gazette* January 12, 1821.

The day after hearing the song, Cass and his party paddled their canoes westward. Although the shore in this direction from Grand Island lacked the same high cliffs as the Pictured Rocks, it was rocky, and for about seven miles they found no suitable place for landing. Finally they arrived at a lovely long beach, known to white men then as Aux Trains and now as Au Train. Here one group of Chippewa and French Canadian explorers who had accompanied the government party from Sault Ste. Marie stopped on the beach to rest before returning to the Sault. Cass, Schoolcraft, and the rest of the main party continued westward.

On the morning the escort party planned to start for the Sault (probably June 24), a strong northern wind swept in from the lake, and great rollers washed over the beach, forcing the men to pull their canoes farther inland to avoid losing them. It was clear that they would not travel that day. By this time the main part of the expedition had reached Point Abbaye, eighty or ninety miles to the west, where they encountered the same storm. Schoolcraft wrote in his journal, "The bay presented a sheet of foam, and our canoes were tossed about with scarcely the power of controlling them. A perfect gale prevailed, and every moment seemed to add to its violence. The swells broke frequently across our canoes, so that one hand was constantly necessary to bail it out, and we expected them to be broke in two at every succeeding swell. . . . Three out of five canoes turned back, and reached the shore in safety, with some injury to the canoes. The other two, consisting of the Governor's and that under the command of Lieutenant Mackay, to which I was attached, after an exertion which exhausted the strength of every person on board, reached the mouth of Portage river, and encamped upon the beach before sun down."

The small group back at Aux Trains was much more cautious, perhaps because the travelers were more familiar with the perils of the lake. The Chippewa told the white men that the evil spirit of Lake Superior, Matchi Manito, was angry. Matchi Manito was a giant water snake with a horned head and a long jagged tail that thrashed about in the deep waters of the lake, inundating the rocks on shore with great waves. To appease the serpent, the Chippewa suggested throwing it an animal to eat. The white men went to a nearby Chippewa village and traded a small knife for an old dog. The Chippewa guides killed the dog, threw it into the lake from a low cliff, and shouted to the lake, "Be quiet!" When the waters continued to roil, the Chippewa announced that the lake would probably not quiet down for another day or two.

The men had set up camp at the eastern end of the beach, where the sand meets low rocks. Nearby a tiny brook flows into the lake. Following the brook a hundred yards or so inland, they crossed an old trail used by the Chippewa to go between Grand Island and Little Bay de Noc. On the other side of the trail, still in sight of Lake Superior, they found an attractive small waterfall cascading over a cliff about twelve feet high. The water arced out from the cliff in such a way that a man could crawl behind it and stand under the fall without getting wet. The water, though cold, was warmer than that of Lake Superior. It was a perfect place to bathe, and all the travelers took advantage of the opportunity to do so, and to rest.

Over the next several days the turbulence persisted, and the men had time on their hands. As they sat around the fire they talked about their recent adventures. All were absorbed with the song of Powers of the Air, the tragic tale in which the reluctant warriors of Grand Island had, in the end, displayed more bravery than any of their more warlike mainland tribesmen. Powers of the Air had personally made a deep impression on them. They were struck by his interest in preserving the story of the heroism of his father and his father's comrades, not in extolling any feats he accomplished himself.

One of the French trappers, the man who had gone to Trout Bay to fetch Powers of the Air for the governor, went down to the beach and walked eastward to its absolute end. There he faced a rock cliff about twelve feet high. He rolled a rock the size of a stool up to the base of the cliff, stood on the rock, and began carving into the cliff, working slowly and carefully with his hatchet and a chisel. The other men gathered around him to watch. During that day and part of the next, gradually a human face emerged and, finally, the ends of two eagle feathers beyond the right cheek. Then the travelers knew: it was Powers of the Air. When the party left the next day, the waters of the lake having at last calmed, all felt that they had left a suitable historical marker.

CREATING
HIAWATHA

AFTER HIS FIRST visit to Grand Island, Henry Schoolcraft became much more interested in native folklore than in mineralogy,
his original field. In 1822 his friend Lewis Cass made him Indian
agent for the Upper Great Lakes with headquarters in Sault Ste.
Marie. There he began to study the Chippewa language and joined
several expeditions westward. During one in 1832 he was the first
white man to see the source of the Mississippi River, Lake Itasca. At
Sault Ste. Marie he became a friend of the Johnston family, whom
he had first met in 1820. John Johnston, an Irishman, was a prosperous trader in Sault Ste. Marie who married the daughter of a famous Chippewa chief. It was she, Ozhawguscodaywaquay, Green
Meadow Woman, who had prevented the military clash with Cass's
expedition. Her English name was Susan Johnston. John and Susan
Johnston's children received educations that combined Western
and Native American traditions: they studied the classics of Greece
and Rome at home and in private schools in Canada while also
learning the language and customs of the Chippewa people. Henry

Schoolcraft fell in love with and married one of these Johnston children, Jane (Obahbahmwawazhegoqua, Star Music Woman), and together they began collecting the stories of the Chippewa. Schoolcraft talked to many local Chippewa about their legends and culture, including the apostate Grand Islander Shingwauk, the chief who had almost attacked Cass and Schoolcraft in 1820. It was Shingwauk who told Schoolcraft about the existence of secret drawings on the shores of Lake Superior.

In 1832 Schoolcraft published two volumes of collected Chippewa tales under the name *Algic Researches*, and in 1851 he began publication of his monumental six-volume series on the history of the native tribes of the United States. These volumes gave only scanty coverage to the visit to Grand Island in 1820. But Schoolcraft did mention "the lonely warrior" and "the gallant thirteen," and he recounted in detail several of the legends that Powers of the Air had told him on Grand Island, including "Mishosha, the Magician of the Lake," and "Kaubina and the Charmed Arrow."

Although Schoolcraft professed great admiration for Indian culture, in his practical work as Indian agent he fully supported the harsh policies of the American government toward native tribes. During the first half of the nineteenth century those policies were based on the idea that the pagan culture of the Indians was incompatible with Christianity and "civilization," and that therefore the Indians in the eastern half of the United States should be removed from their ancestral homes and sent west. The assumption underlying this policy was that Indians were an inferior race, biologically and culturally. According to President Andrew Jackson in 1835, "All preceding experiments for the improvement of the Indians have failed. It seems now to be an established fact that they cannot live in contact with a civilized community and prosper."

In 1836 Henry Schoolcraft and Lewis Cass, who had become secretary of war and head of the Indian Department, negotiated the Ottawa-Chippewa Treaty of Washington, a sweeping agreement that ceded thirteen million acres of land to the United States, including almost all of the Upper Peninsula east of a line between Bay de Noc and Marquette. With the exception of several tiny reservations, all the lands belonging to the Grand Island, Bay de Noc, Mackinac, and Sault bands of Chippewa passed to the whites. Yet the treaty included a provision that the Chippewa could continue to fish and hunt these lands until they were required for settlement. By the time those settlements were established, much of the suitable land in the western United States had already been claimed by others, so the Chippewa were never forcibly sent west. Hard as their lot was, they were not subjected to removal and to the "Trail of Tears" experienced by the Cherokees and other eastern Native Americans. But they retained only a few thousand acres of their land, and within a few decades they were reduced to poverty.

In negotiating the 1836 treaty Schoolcraft made full use of his kinship ties to the Chippewa. At the same time that he represented the U.S. government, his wife's uncle and other relatives represented the Sault Ste. Marie band with whom he negotiated. In pursuing the government's goals Schoolcraft thus had enormous advantages. What is more, Schoolcraft and Cass plied the Chippewa chiefs with liquor to soften their resistance, even though both had often spoken against such practices.

In their attitudes Schoolcraft and Cass were typical of white politicians in the first half of the nineteenth century. On the one hand, they expressed sentimental views about the disappearance of the "noble red man" and actually did valuable research in collecting information about Native American culture; on the other, they were convinced that the native culture was doomed by the advance of "civilization," and energetically and unfeelingly promoted the destruction of that culture.

In Cambridge, Massachusetts, several Harvard faculty members were interested in Lake Superior and the area around it, now rapidly opening up to exploration and settlement. In 1848 Louis Agassiz, a new professor of zoology who had already achieved recognition in his native Switzerland, led an expedition of sixteen men (eleven of whom were faculty or students at Harvard) to study the natural history of Lake Superior. When they returned to Cambridge they told fascinating tales to their friends and colleagues about the beauties of the area. In 1850 Agassiz published one of the first scientific examinations of the lake, filled with data supporting a creationist theory of natural development. When the naturalist looks at Lake Superior, he wrote, "He beholds indeed the work of a being *thinking* like himself, but he feels at the same time that he stands as much below the Supreme Intelligence in wisdom, power and goodness, as the works of art are inferior to the works of nature." (Agassiz' creationism became better known after 1859, when he opposed it to the form of evolution presented by Charles Darwin in *On the Origin of Species.*)

Agassiz told his close friend and colleague Henry Wadsworth Longfellow, a specialist in European literature, about the expedition's adventures on the great lake. Longfellow, who at that time was interested in the folklore of the American Indian, found the writings of Henry Schoolcraft particularly provocative. He wrote, "I pored over Mr. Schoolcraft's writings nearly three years, before I resolved to appropriate something of them to my own use." In 1854 Longfellow began an epic poem on the American Indian, which he first called *Manabozho* but later changed to the more euphonious *Hiawatha,* whose name occurred more commonly among the Iroquois. Longfellow stated, "The scene of the poem is among the Ojibways on the southern shore of Lake Superior, in the region between the Pictured Rocks and Grand Sable." This is the area of Lake Superior shoreline that stretched before Little Duck on the final lap of his daily runs along Trout Bay beach, when he spied Grand Portal emerging beyond Trout Point.

Longfellow drew upon elements of many of the legends of Lake Superior preserved by Schoolcraft. He included Mishosha's magic canoe and one of the secret words for speeding it up that Powers of the Air had revealed to Henry Schoolcraft, although he spelled it *cheemaun* instead of *chemaun,* as Schoolcraft had recorded it:

> *Straightway then my Hiawatha*
> *Armed himself with all his war-gear,*
> *Launched his birch-canoe for sailing;*
> *With his palm its sides he patted,*
> *Said with glee, "Cheemaun, my darling,*
> *O my Birch-canoe! leap forward,*

Where you see the fiery serpents,
Where you see the black pitch-water!"
  Forward leaped Cheemaun exulting,
And the noble Hiawatha
Sang his war-song wild and woeful.

Longfellow's poem was a tremendous success. Four thousand copies were sold on publication day, and two months later it was still selling at the rate of three hundred a day in Boston alone. Schoolchildren all across America would memorize it, some voluntarily, others involuntarily. Abroad it became known as *the* poem of the American Indian. Inevitably its overwhelming success soon brought a train of critics: some said it was saccharine, borrowed from other sources, or colored by Longfellow's and Schoolcraft's prejudices. No one could deny its immense popular appeal. It has been loved by millions of people.

SETTLING THE ISLAND

ON JULY 30, 1840, at the invitation of Omonomonee (also called Monomonee), one of the last chiefs to summer there, the first permanent white settlers came to Grand Island. Abraham Williams was a farmer originally from Vermont who had temporarily settled in Decatur, Illinois. He arrived at Grand Island with his family on the schooner *Mary Elizabeth* from Sault Ste. Marie. He had considered making his home in the Sault, but after seeing all the drunkenness and violence there he decided that it was no place to raise his children. He was looking for land where he could farm and run a store.

Williams was an industrious, talented, and honest man who had learned many skills in Vermont and Illinois. He was a blacksmith, cooper, boatbuilder, tanner, carpenter, and merchant. He immediately claimed four of the log cabins left by the American Fur Company on the south end of the island, and he quickly began building chicken coops and pigpens, a blacksmith shop, and, eventually, several houses. Some of those houses are still standing today.

Powers of the Air, along with other mainland Chippewa, still lived on the island during the summers. Williams hired several of them to help him with his work, and he treated them fairly, much better than the agents of the American Fur Company had done. He soon learned the story of the original Grand Island Chippewa, and he was especially interested in Powers of the Air, whom he employed on a permanent basis for several summers.

By this time lake boats were appearing ever more frequently, a few of them steam-powered. They usually stopped at Grand Island to take on fresh supplies, especially wood for fuel, and the owners of these vessels sold manufactured goods to Williams for his store. Williams was educated, and he sought newspapers and books as well as cooking utensils, tools, guns and ammunition, and other items from eastern factories. He traded furs, fish, maple syrup, berries, and deer meat for these goods. Later he had milk, cheese, butter, pork, and beef for sale.

Powers of the Air became a family friend to Abraham and Anna Williams and their children. Working every day with the Williams family for several years, and with other white men in subsequent years, Powers of the Air became fluent in spoken English.

He was impressed with the English language. He noticed that when Williams read newspapers and books he would sometimes burst into laughter or become quite irritated. Clearly, this language possessed magical powers. Anna Williams offered to teach him to read. But he was already over forty years old when the white settlers arrived, and learning to read proved to be too difficult, although he did learn the alphabet and a few words. For a while he wore a deerskin coat onto which he had sewn the letters of the English alphabet, some of them in incorrect or distorted forms.

Williams came to trust Powers of the Air and noticed that unlike many other Indians, he did not abuse alcohol. For a while Williams asked him to manage his store, but gradually he realized that without reading ability or knowledge of arithmetic the task was too much for him. Powers of the Air could handle things so long as they remained on a barter basis or if the numbers involved did not go over twenty or thirty. For elementary calculations he kept a bag full of small stones that he used like an abacus without supporting rods. But as the business became complicated, Williams had to step in more and more often.

There was plenty of work for Powers of the Air outside the store—building fences, gathering maple syrup, feeding the animals, bringing in stones from the quarry by horse and wagon, and acting as a general handyman. He taught Williams where to catch the best fish and how to make many simple items from materials found on the island—deerskins, cedar bark, and birchwood. Williams taught him how to forge iron, build barrels, and make fireplaces.

Powers of the Air continued to live on the island in the summer, as did some other Chippewa. He stopped talking about his father and the Battle of the Cavern, since few were interested. The island Chippewa had their own chief, first Omonomonee and later Kishkitawage, Man with an Ear Cut Off, both of whom traced their authority and traditions to different sources from the ones Powers of the Air had known as a boy. For him to emphasize that he was the son of the old chief of Grand Island would have caused trouble among the few islanders remaining.

CUNNING
VERSUS WAR

IN THE SUMMER of 1843, three years after the Williams family arrived, the biggest Chippewa war party that anyone had seen for years arrived at the Williams landing on the island. There were more than five hundred warriors, most of them from Leech Lake, Minnesota, but also about thirty from Ontonagan. They had just completed a successful raid against the Sioux to the west, and now they were going to Sault Ste. Marie, where they hoped to replenish their ammunition and supplies and to receive yearly rations from the U.S. government. They were in a boisterous mood and were looking for a way to celebrate their victory. One of their chiefs came to Williams' store, where he was greeted by Powers of the Air.

The conversation was friendly, but there was not immediate agreement on a number of items. Powers of the Air expressed amazement that the Chippewa were still fighting the Sioux. "It will gain you nothing," he said. "My own Chippewa on the island made the mistake, only once, of fighting a war, and we were almost all killed. My band of Chippewa has disappeared."

"But we have avenged you," said the chief. "We took many Sioux scalps and won a great victory. I have won the right to call myself *'ogitchida,'* great hero."

"You are fighting your brothers at a time when the white men are taking over the entire country," said Powers of the Air. "We do almost as much damage to ourselves as they do to us. I notice that almost all of your men have guns. That means that when you fought the Sioux you were able to kill more of them than in the old days, when our weapons were weaker. Meanwhile the white men grow stronger and stronger. Mr. Williams has shown me in his books and newspapers pictures of their latest weapons. They have big guns, cannons, as large as your canoes. They have ships with black metal engines that drive them forward. Mr. Williams says that they now have iron horses that run on tracks and can pull many wagons filled with supplies. Soon they will take all our lands. They are far more numerous than we are."

The chief objected, "If they are so powerful, what good would it do for us to stop fighting each other and fight them? They would defeat us."

"We should not fight them or anyone else," replied Powers of the Air. "That was the way the Grand Island Chippewa once lived. But we should preserve our strength, claim our lands, keep our culture, and educate our children, both in our traditions and in the new skills they will need. We will be able to preserve some part of our lives that way."

"You do not speak like a Chippewa," said the chief. "Our fathers and grandfathers have always fought the Sioux. You have been around the white men too long. Now let us do some trading. We want whiskey to celebrate our victory. We will pay you with captured weapons from the Sioux."

"Mr. Williams does not like to sell whiskey to Indians," said Powers of the Air. "He may not have much. He rarely drinks himself. I will have to talk to him to find out if he wishes to sell you any whiskey." Powers of the Air had spoken the truth, but he also knew that Williams had a considerable cache of whiskey hidden away for use in trading.

Powers of the Air went to talk to Williams, who had been watching the conversation from near his house, where he had asked his wife and children to stay inside, out of sight. Powers of the Air relayed the chief's request and added, "If you sell them whiskey, they will immediately drink it, and perhaps then, after they have lost their senses, do you harm. If you tell them you have no whiskey, they will not believe you, and perhaps forcibly search your entire house and all other buildings for it. And if they find it, out of irritation with you they will not pay and may do much damage after they are senseless with the drink. There are so many of them that it would be hopeless to resist."

Williams listened and then said, "Tell them that they have just arrived and must be tired and hungry. We will give them some food tonight, and they can eat and rest. Tomorrow night they can have a much better celebration, and I will find some whiskey somewhere."

Powers of the Air relayed this message to the chief, who seemed ready to accept it. But his men were unhappy and stayed awake much of the night.

More than sixty years later Anna Williams, who had been fifteen at the time, told a woman from Marquette about that evening. Anna recounted, "They danced and powwowed all night! Yelled

and beat their drums, and sang their war songs! We sat up all night to bake bread for them. Yes, we *were* kind of nervous. Of course they wanted whiskey, but Father wouldn't give them any that night. He was afraid to."

The next day Williams told the Chippewa chief that he would give them whiskey free, if they would follow his instructions. He would row in his boat to a nearby place with the whiskey and then distribute it to all the chiefs, twelve in number, who in turn, after Williams had left, would give the whiskey to their men. The Chippewa agreed.

Williams loaded several small barrels of whiskey into a wooden boat and rowed across Murray Bay to Agate Point, the southwest portion of the Thumb. There he grounded his boat and placed the whiskey on the stony beach. He immediately reembarked and returned home, not waiting to see the results.

From his home he could barely see the Chippewa on the spit of land. They were merely tiny figures surrounding their large war canoes. The canoes and figures were there for three days. Then they disappeared to the east, headed for Sault Ste. Marie.

SHOOTING
THE TOTEM

IN THE 1850s and 1860s many Chippewa in the Grand Island/Pictured Rocks area moved away from their traditional homes. Dispossessed of their hunting and fishing areas by treaties and by white settlements, they looked elsewhere for places where they could survive. Some went to such growing towns as Marquette, St. Ignace, Escanaba, and Sault Ste. Marie, usually settling on the edges of the communities in small groups of miserable huts. Others moved to the tiny reservations left to them by the treaties of 1836 and 1855. Several Chippewa who had traditionally spent part of the year on Grand Island moved to the reservation at Bay Mills, not far from Sault Ste. Marie, where their graves can still be seen in the cemetery. Not far from the cemetery lives today the Menominee family, descendants of Omonomonee, one of the last chiefs to summer on Grand Island.

The cemetery itself reflects the massive changes occurring in the Chippewa culture during these decades. The older graves are the traditional low wooden spirit-houses, with no markers except for an occasional surviving totem sign. The deceased inhabitants of these houses are not identifiable by name to the casual observer, although the local Chippewa still sometimes know. Many of the graves dating from after about 1860 or 1870 have headstones in the white style on which the names are engraved. But the inhabitants of these graves suffer from another kind of anonymity, since in many cases their names changed during their lives, from their original Chippewa ones to English ones, sometimes quite nondescript, such as Clark or Smith. Thus one tombstone memorializes Martha Smith, born on Grand Island in 1832, who died at Point Iroquois near Bay Mills in 1879. What was Martha Smith's original name? Perhaps some of her descendants know. Did *Smith* take the place of a name similar to those of her earlier fellow Grand Island women, such as Sound of Wind in the Trees, Little Red Flower, or Pearl-Whisper?

With the end of the indigenous Grand Island culture by the middle of the nineteenth century, the old Chippewa burial ground near Cemetery Beach was left unattended. Not long after white settlers built a small village on the mainland, later known as Old Munising (a mile or two northeast of present-day Munising), venturers began to explore the island. One of them, a man who in the late 1860s or early 1870s built the first hotel and saloon in Old

Munising, "discovered" in the old Indian cemetery on Grand Island a curious "totem pole," covered with figures of birds and animals and faded patches of red. Deciding that it would make a fine decoration for the entrance to his business establishment, he installed it next to the front door of the hotel. Everyone agreed that it enhanced his watering hole. Hunters and lumberjacks began to call the place the Bird-Pole Saloon.

The pole stood there until one afternoon about 1880, when a group of hunters in the saloon who had drunk more than was good for them began to speculate about it. "What is it, and where did it come from?" one of them asked. Nobody really knew. "It is an Indian idol," said the bartender, but he had no idea of its origin. "Let's see how that pagan idol stands up to a little target practice!" cried one of the hunters. He went out into the street, pointed his gun at the pole, and started firing away. The other hunters soon joined him, and within a few minutes the pole was shot to pieces.

The next morning a few residents of the town noticed the vandalism and expressed regret at losing such an interesting marker. The story of the pole was told around town for a while, although no one knew what the pole had represented.

SHINING THE LIGHT

IN THE MIDDLE of the nineteenth century U.S. business interests decided that the only way to exploit the mineral riches of the Lake Superior region was to construct a deep canal at the Sault rapids on St. Mary's River connecting Lake Superior and Lake Huron. A group of eastern investors formed a private company, the St. Mary's Falls Ship Canal, to construct a navigable channel between the two lakes, and in June 1853 a crew of four hundred men under the direction of Charles T. Harvey began digging the ditch. The path of the canal ran right through the traditional village, fishing, and burial grounds of the local Chippewa. Notwithstanding their protests, the canal was rapidly built, and in May 1855 the locks at Sault Ste. Marie were opened. For the first time Lake Superior became accessible to large ships.

To guide ships along the dangerous and rocky coasts of Lake Superior the U.S. government constructed lighthouses on its prominent capes, points, and islands. In 1855 a team of men from the Light-House Establishment landed on the northern point of Grand Island and erected a lighthouse several hundred feet above the lake atop a sheer cliff. Officially known as Grand Island Light

Station, it was called North Light by the island's inhabitants and visitors. The lighthouse was located on the most remote spot of the island. Access by water from the nearby mainland involved several hours of sailing beside the cliffs of the island exposed to the strong north winds of Lake Superior. Access by land involved travel by boat to the southern end of the island, then a ten-mile hike along a forest trail. The trail was the same one used by the original Grand Island Chippewa to go to the site of the Stone Pot Feasts.

One of the early keepers of North Light was William Cameron. Cameron was born north of Lake Superior, in Canada, and educated in Toronto in European languages and literature. A career as a scholar did not materialize, and eventually he moved to the United States, settling in Sault Ste. Marie, Michigan. There he became fascinated with Chippewa culture and, according to his white acquaintances, "went native." He learned the Chippewa language, married a young Chippewa woman named Sophie Nolan, and raised a family that eventually included eleven children. William and Sophie always spoke Chippewa to each other, and all their children grew up speaking Chippewa. However, William retained his love of European literature and accumulated a large library of books in several languages. He gave his children lessons at home and required them to learn both Chippewa and English.

When Cameron came to Grand Island as keeper of North Light he installed his library on shelves lining the main hall of the lighthouse, leading up to the tower. As the sole keeper of the lighthouse, he could arrange his schedule as he saw fit, finding time for reading and writing as well as fulfilling his keeper's duties. But since he was up all night tending the light, and since he also had to hunt and fish in order to get food, the hours he could devote to study were limited.

The light was a beautiful instrument, a fourth-order Fresnel lens consisting of many prisms assembled in a bee-hive shape. The lens was manufactured in France, and on its base was a shiny brass plate engraved with the name of its maker, "L. Sautter, Paris." It was a fragile affair that required considerable care. Sunlight was never allowed to strike it, since the sun's rays could be refracted and magnified by the prisms in such a way that the heat could warp the glass or cause a fire. At dawn Cameron always extinguished the light and drew the protective curtains in the lightroom.

At night the lens revolved on clockwork machinery that caused the light to flash every ninety seconds, a pattern unique to North Light so that ship captains would know where they were when they saw it. The machinery was powered by pendulum weights attached to cords that descended from a brass drum in the lightroom through pulleys and then over brass sheaves into flues that extended the full length of the tower between its double brick walls.

William and Sophie Cameron soon came to the attention of Powers of the Air, who, now in his sixties, still spent his summers on the island. The three became close friends. William Cameron spent hours with Powers of the Air discussing the history of the island. Cameron attempted to write poetry in Chippewa, a serious test of his linguistic skill. (One of his poems in the language was passed down to his granddaughter Julia, who recited it to me when I visited her in Au Train for the last time in 1985.) Powers of the Air was interested in language too, especially the English language, and he asked Cameron to read aloud to him from his books on English literature.

Powers of the Air saw that Cameron did not get enough sleep because of the need to tend the light all night long, and he offered his help. The management of the light was a routine that was strictly defined in the regulations of the Light-House Establishment, which specified that no unqualified person was allowed to attend the light. Nevertheless, Cameron taught Powers of the Air how to trim the wick of the light, adjust its height, replace the delicate mantle when necessary, and wind up the weights in the tower wall so that they drove the clockwork properly, delivering the ninety-second signal. Powers of the Air then took turns tending the light, alternating with Cameron so that both of them could get some sleep during the night.

In the mornings Powers of the Air and Cameron would often hunt together, seeking partridge, rabbits, and occasionally a deer.

NORTH LIGHT

Both had rifles, but Cameron was not an experienced hunter. Powers of the Air taught him where to find game and how to hunt most effectively. In the afternoons, after they had dressed their kill (if they had been fortunate that day), Cameron would sit in the living room of the lighthouse and read aloud from his books. Although Powers of the Air never mastered reading, he learned much about literature from Cameron and acquired a remarkably rich English vocabulary.

The lighthouses of Lake Superior were supplied by lighthouse tenders, ships that also brought uninvited inspectors. In the 1870s the tender serving the lights of Lake Superior was the *Dahlia,* captained by John Hallaran, master. Captain Hallaran supplied the lightkeepers not only with canned food, barrels of flour and sugar, lard oil for the burners in the Fresnel lamps, kerosene for reading lamps, and coal for the stoves but also with a rare luxury, a circulating "Lighthouse Library." Each allotment of books was delivered in a special case, with an inventory of its contents on a sheet attached to the underside of the lid. To obtain a new case of books when the tender made its next visit, the keeper had to return the old case, and the petty officer from the tender carefully checked each of the books against the inventory to make sure that none was missing.

Cameron was critical of the books that came to him in this way. Most of them were adventure novels designed to provide some entertainment for lonely lighthouse keepers, many of whom were barely literate. He begged Captain Hallaran to obtain for him some recent works of literature, especially poetry. Fulfilling this request was not easy, since all the books supplied to the tenders for distribution to the lights came from inventories approved by officials in the Light-House Establishment. But Captain Hallaran was able to persuade his superiors in Detroit to ask for a greater variety of books to be shipped from Washington, and he began setting aside all poetry for William Cameron.

One day the *Dahlia* delivered to North Light a case of books containing Henry Wadsworth Longfellow's *The Song of Hiawatha.* William Cameron already knew about Longfellow and had read his famous poem concerning the Chippewa on Lake Superior. He had, in fact, already told Powers of the Air about it, and had recited a few lines from memory. As soon as Cameron saw the book he knew that he must read it to his wife and to Powers of the Air. He told them that he had a treat in store for them, a long poem that was about their people and their land.

Cameron began reading late in the afternoon of a languid summer day. All three of them sat at the dining table in the living room of the lighthouse, where they could look out the windows to the cliff edge and see the waves coming in to the rocks in jagged rows, several hundred feet below. At that time of the year, darkness would not come for another five or six hours, so they had a lot of time before William would ascend the cast-iron spiral staircase and light the lamp in the lantern room.

William Cameron began reading the words of *The Song of Hiawatha*, which by this time were already familiar to millions of people. Powers of the Air listened attentively. His eyes glistened a bit at Longfellow's mention of the Ojibways, the great bear Mishe-Mokwa, and the sturgeon Nahma. And he smiled when he heard the lines that Cameron had earlier recited from memory:

> *By the shores of Gitche Gumee*
> *By the shining Big-Sea-Water*
> *Stood the wigwam of Nokomis,*
> *Daughter of the Moon, Nokomis.*
> *Dark behind it rose the forest,*
> *Rose the black and gloomy pine-trees,*
> *Rose the firs with cones upon them;*
> *Bright before it beat the water,*
> *Beat the clear and sunny water,*
> *Beat the shining Big-Sea-Water.*

But to much of the poem Powers of the Air and Sophie showed little or no reaction. They listened politely, but it was clear they found the poem foreign, especially in its rhythm and rhyme. As the story progressed, occasionally one of them would smile or frown, as at something that seemed a bit familiar but not quite right. At one point Powers of the Air interjected, "This is not Hiawatha, but it may be Manabozho, and he was never so solemn. He was always playing tricks and getting either himself or others into trouble."

After a while Powers of the Air and Sophie began to catch on to the exotic but repetitive cadence of the poem, anticipating its beat and taking pleasure in its description of the forests and shores of Lake Superior. After more than an hour of reading, William Cameron came to the section of the poem in which Longfellow had relied on the story of the magic canoe of Grand Island, the spot where Hiawatha

> *Launched his birch-canoe for sailing;*
> *With his palm its sides he patted,*
> *Said with glee, "Cheemaun, my darling,*
> *O my Birch-canoe! Leap forward."*

On hearing the word *cheemaun* come from Cameron's lips, Powers of the Air leaped up from his chair, spilled William's coffee as he knocked into the table, raised his arm in the air, and cried out, "He got it from me! I am the only person who ever told that secret!"

"How could he have got it from you?" objected William. "Did Longfellow ever come to Grand Island?"

"No, but the secret words were revealed to white men only one time, and that was when I told them to Henry Schoolcraft when he came here many years ago. I made this mistake because all our men had recently been killed, and I was very discouraged. I thought it made no difference."

"Longfellow said that he studied the legends of the Chippewa very carefully before he wrote this poem," responded Cameron, "and Schoolcraft recorded the Chippewa legends more fully than any other person. He must have written these words down in one of his books, and Longfellow found them there."

Cameron turned to the introductory material in the book, and there, indeed, he found Longfellow's statement of gratitude to Henry Rowe Schoolcraft, who had provided material "appropriate for my own use." Thus it was clear that parts of the legends of Grand Island had now come full circle.

All keepers of lighthouses were required to keep logs in which they recorded daily weather observations, ship sightings, and, of course, any indications of ships in distress. William Cameron dutifully fulfilled all these requirements, and in addition he inscribed in the official log his personal reaction to all events. Early one August morning Cameron wrote:

> At 2:30 a.m. a large eared owl sat perched upon the railing of the tower, as sedate and important as a Judge Advocate upon a court martial. But when the bull's eye of the Fresnel lens would flash upon him, he would throw up his wings and cast his head down, as much as to say, "I submit." But no sooner would the flash be off of him, when he would hoot—as if he called "Come on McDuff!" He finally retired to his native forest where he came upon a "brither chum" with whom he had quite a conversation. Although I am an old resident in the woods I never could learn their language thoroughly, which prevents me from committing his honor's discourse to paper.

On May 8, 1874, as Cameron watched the snowdrifts disappear after a long winter, he rhapsodized:

> *Lo! Spring unlocks her many frozen lands*
> *And melts the icy jewels from her hands;*
> *From long repose she wakes the sleeping flowers,*
> *Whose fragrant smiles make glad the passing hours;*
> *Her verdant carpets all the fields adorn,*
> *The dew-decked forest sparkles in the morn;*
> *The feathered songsters tune their gladsome lay,*
> *While genial showers and sun-shine guild the day.*

Two weeks later he sighted in the lake the first ship he had seen in six months and exuberantly wrote in the log: "Hip, hip hurrah! A propeller rounds the point, on her way up to Marquette, ploughing through the ice."

The year 1875 was a fateful one for Cameron. Sophie's health constantly declined, and he soon realized that she would not recover. Overwhelmed by distress at his wife's condition, he was scarcely able to perform his duties as keeper of the light. Fortunately, Powers of the Air was there to help. But on the evening of June 30, 1875, the revolving mechanism for the flashing light stopped, and for what seemed like an interminable time the light did not operate properly. The cord that wound around the drum before descending to the weights in the tower walls kept slipping off its track, probably because Cameron had, in his grief, hastily wound up the mechanism, not making sure that the cord was not twisted. Cameron asked Powers of the Air to turn the lamp by hand, counting out loud "one little second, two little second, three little second," trying manually to produce the light's correct timing. While Powers of the Air turned the lens, Cameron unwound the entire cord and then carefully rewound it, eliminating the twists that had caused the problem.

Upon completion of the repairs, Cameron knew that he was obligated to report the infraction in the log. After all, the light had operated irregularly for many minutes. He wrote that he had had a problem with the cord, but had repaired it while "the assistant" turned the lamp by hand. This reference to an assistant was something new in the log. Never before had Cameron acknowledged the presence of Powers of the Air at the lighthouse.

In the weeks following this incident Sophie Cameron's health rapidly deteriorated. Finally, on November 7, 1875, she died, and Cameron fell into complete despair. Powers of the Air ran the light single-handedly for several weeks. Cameron was scarcely able to fill out the daily log entries, but he managed to write a lament to his wife:

> Full many a gem of purest ray serene,
> The dark unfathomed caves of ocean bear,
> Full many a flower is born to blush unseen,
> And waste its fragrance in the desert air.

The following summer the steamer *Dahlia* with Inspector McCann on board made a surprise visit to evaluate North Light. McCann found serious deficiencies in Cameron's performance of duties. He spotted the reference to "the assistant" in the log and inquired about the identity of this unknown person, since there was no approved assistant keeper at North Light. Cameron confessed that a local Chippewa often helped him to maintain the light, stressing that he was remarkably loyal and competent. McCann replied that no assistant was authorized, pointed to the paragraph in the U.S. government's *Instructions to Light-Keepers* stating, "Keepers must not allow visitors to handle the apparatus," and warned Cameron that he had violated the regulations. Upon examining the log further, McCann found many other infractions. He issued a letter of reprimand to Cameron, citing him for "illegally relying on an unauthorized bystander for help in performing his duties, a person who was, furthermore, an illiterate native incapable of understanding the lighthouse establishment regulations." He further chastised Cameron for failure to keep a proper official log of the station, filling it with "personal observations and frivolous poetry."

After this inspection and official censure, Cameron told Powers of the Air that he could no longer help him at the lighthouse. Cameron also no longer entered poetry or personal observations in the official log. A typical entry now was: "July 1, East wind, light and clear." Although Powers of the Air continued to visit from time to time, he no longer spent much time at the light.

SPREADING
THE GOSPEL

IN JUNE 1883 a professor and Baptist minister named Columbus Horatio Hall came from southern Indiana to the Munising area to spend the summer. Interest in the south shore of Lake Superior had been growing for several decades, initially sparked by the publication of Longfellow's *Hiawatha* in 1855. In the 1860s several national publications, including *Harper's New Monthly Magazine*, had carried articles about the Pictured Rocks area. Professor Hall read these publications, and the prospect of an escape from the Indiana heat to the fresh north woods was appealing. A trip to Munising, now possible because of the construction of a railroad there in 1881, would provide welcome relief. Professor Hall was the son of Indiana pioneers and knew how to hunt and camp, and his companion—Judge David D. Banta of Franklin, Indiana—also was an experienced hunter. They expected no trouble surviving in the remote forests. Besides, as an ordained minister Hall looked forward to an opportunity to spread the gospel among the local Chippewa.

In 1883 the present town of Munising did not yet exist. The closest railroad terminal was about five miles south of where the town is today. At the station Professor Hall and Judge Banta hired a wagon and driver to take them to the shore of Lake Superior. There they found a tiny settlement—later known as Old Munising—near Munising Falls, not far from today's Munising hospital. They looked across the bay, saw a large and lovely island, and learned that it was named Grand Island.

Hall told a boy they met that they would like to make some explorations, especially of the Pictured Rocks and Grand Island, and asked if he could help them find a local guide. The youth said that the person who knew the most about the area was an old Indian, named Jim Clark, living up in the hills in Indiantown, several miles to the east. Old Munising had a small livery stable, and Hall hired a horse for the afternoon. Leaving Judge Banta with their belongings on the lake shore, he rode to Indiantown.

Indiantown turned out to be a collection of some ten or twelve shacks. The minister-professor was shocked to see the poverty of the place. Most of the dwellings were nondescript piles of boards, rugs, and pieces of tin. Each was assembled in its own distinct way, taking advantage of any castoff building materials available at the moment of construction. All were dirty, shabby, and ramshackle.

The only person visible was an Indian woman carrying a dented bucket to the well that obviously served the entire village. Hall rode up to her and stopped. "Excuse me," he queried, "but do you know a man named Jim Clark? Does he live here?"

"I suppose you mean old Jim," she replied; "young Jim lives in town. Of course I know where old Jim lives. He's right there." She pointed at a pile of boards across the path that Hall decided, despite appearances, had to be Jim Clark's house.

It was more like a large decaying mound than a house: a disorderly arrangement of boards of varying lengths set upright in the ground in a rough circle, surmounted by pieces of worn and variegated roll roofing of various colors forming an uneven dome. The house looked like an upside-down basket, with torn sides and bottom. Rocks and short logs were dispersed over the roof to keep the wind from picking up parts of it. From the center of the roof, at the top of the dome, a corroded black stovepipe protruded, and from the stovepipe dark gray smoke streaked almost horizontally to the ground in the robust cold wind coming off the lake. In front were two stakes stuck upright into the ground with several branches connecting them. Hanging from the branches, high enough to be out of reach of small animals, were several drying fish. Discarded utensils and pieces of equipment were scattered around the front yard: a large, old, mottled gray coffeepot with no bottom, a small plow with the handles missing, and the frame of an old canoe that had not been usable for years. Everything was decrepit, with one exception: leaning against the wall near the entrance was a pair of snowshoes in excellent condition.

The entryway seemed to be several layers of old rugs, but beneath them Hall found a wooden door with a bona fide knob. The rugs had been placed there for insulation during the winter and had not yet been removed for the summer. Hall knocked on the door and shouted, "Anyone home?" He was pretty sure, from seeing the smoke, that the place was inhabited.

After a while the door opened and an old Chippewa man emerged, stepping out into the sunlight. He was tall, but his top half was bent over as he came through the low door, and when he straightened up the effort was only a partial success. Hall guessed that he was in his eighties and immediately doubted that he could serve, at his age, as a proper guide. He was dressed in dark and soiled trousers, held up by a rope through the belt loops, and a frayed woolen hunting shirt with yellow underwear peeking from the collar. On his feet were curious, obviously homemade, shoes; the tops looked like traditional Indian moccasins but were undecorated except for a distinctive puckered lengthwise seam. On the bottoms were black soles evidently made of several layers of felt roofing somehow attached to the tops. The old Chippewa had tarred the roofing and sprinkled sand on the bottoms before the tar hardened. The result was a waterproof shoe bottom. His face and neck displayed deep gullies. He was not exactly filthy, but he was not clean either.

"Are you Jim Clark?" Hall asked.

"Yes, I am. Who are you?"

"I am Reverend Columbus Horatio Hall, and I am looking for a guide around the Pictured Rocks and Grand Island. Some people told me that you might be able to help."

"Why did you come here?" Clark wanted to know.

"I came because I heard it was a beautiful area and also because I would like to preach God's word to the Indians."

The old Chippewa man smiled and said, "You are a bit too late. You know, we Indians have our own church, and in fact I am the lay minister for our congregation. What church do you belong to?"

Hall answered that he was a Baptist.

"Well," observed Clark, "I am a Methodist, but we still have something in common—we are both ministers." And he stuck out his hand, which Hall vigorously shook.

"Do you know your way around Grand Island?" Hall inquired.

"Well, yes, I've been around it all my life."

"Would you, or someone you suggest, be willing to show me around the place?" queried Hall, fearing that at his age Jim Clark might not be up to a strenuous trip.

"I think that it might be possible," replied Clark, "especially if you would be willing to add something to the collection plate of our church."

"I think that can be arranged," Hall agreed.

"It's too late to start out for the island today," observed Clark, "and this wind is a bit stiff. Why don't you return to town and arrange with the keeper of the store there, Mr. Cox, to rent a sailboat? We will go tomorrow, if the wind dies down a bit, as I think it will. Mr. Cox has a boat which he rents out. Tomorrow morning I will walk to town and meet you at the dock where Cox keeps the boat. But if the lake is rough, I'll not come, and we will wait until the following day."

"That's a great idea," exulted the professor. "I love to sail. I used to do it at the university when I was a student. But don't walk to town. I'll send the wagon that brought me from the railroad station. And I'll use the wagon to pick up supplies for all three of us—you, Judge Banta, and myself."

"You won't need a wagon to pick up supplies. There is only one

store in town, and it's right at the dock. But I heard that Mr. Cox is
not feeling well, so he may not be open. He lives over the store, and
I am sure that he will rent his boat even if the store is closed. If you
have a little food with you, we will be fine at this time of year. We
can find some more on the island. There is good fishing there. Do
you have fishing equipment and a gun? Are you a good shot?"

"Both Judge Banta and I have fishing gear, rifles, and a shotgun,
and we are experienced fishermen and hunters."

"Fine, I'll see you tomorrow morning," rejoined Clark.

Hall returned to town, where he found, as Clark had predicted,
that the store was closed, but its proprietor, William Cox, was willing to rent his boat. The professor and the judge had in their bags
enough food for several days.

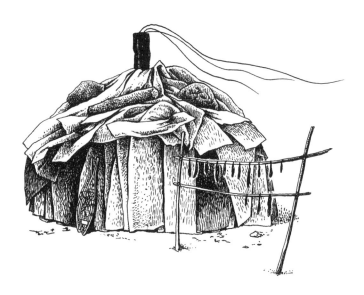

RETURNING TO TROUT BAY

THE NEXT DAY was fairly calm, with a soft breeze from the south. The water was glass-smooth near the pier and patterned with light wind streaks farther out in the bay. No whitecaps were visible. It was a perfect day for an unhurried sail.

Early in the morning Columbus Hall sent the wagon out for Jim Clark. In about an hour the wagon returned. Clark was carrying a jacket and a small bag but no gun. Hall introduced Clark to Banta, and then the three men loaded the professor's and the judge's supplies, five or six large packs, including a large tent, a small metal stove with a detachable oven, and several guns. As Clark helped stow the gear it was clear that although he moved slowly he could get around fairly well.

As soon as the boat was twenty feet from the pier and Hall adjusted the small sail, it caught the tender breeze, and the vessel moved out slowly into the bay. At the leisurely rate of the morning wind their trip was clearly going to last an hour or more. In the distance the island loomed, the morning sun catching the cliffs on the east side of the Thumb.

To Hall's surprise, Clark began to ask him questions. "What do you do for a living?" was his first query. "Nobody I know can make a living as a minister."

"I teach the Greek and Latin languages in a small school in southern Indiana named Franklin College," Hall replied.

"I did not know that people in Indiana speak Greek and Latin," Clark observed.

"Oh, they don't, but we believe that knowledge of the classical languages, even if you do not speak them, is necessary for an under-standing of our culture, which is based on Greece and Rome."

"So you do not speak these languages, but people do speak them in Greece and Rome?" Clark queried.

"Classical Greek and Latin are not spoken today anywhere, even though much great literature is written in them."

"You teach languages that no one speaks, while up here no one teaches a language that quite a few people still speak, the Ojibway or Chippewa language," Clark observed.

Hall admitted that this was too bad but said that to the best of his knowledge, Chippewa was not a written language, and there-fore there was no body of literature to study.

Clark agreed, adding, "But we do have some legends and stories."

Hall said that he would like to hear them sometime.

Clark was silent for quite a while, staring distractedly at the in-distinct wake left by the boat. Banta was humming a song and hap-pily taking in the scenery. Finally Clark said to Hall, "It is good that you are so interested in the old stories of your people, studying them in detail and preserving them well. Unfortunately, no one has done that for the old stories of my people, and now it is too late. I am one of the oldest Chippewa in this area, and yet I remember only a small portion of our legends. I was not one of those few of our ancestors like Nahbenayash, Line of Thunder Clouds, who made an effort to commit many of them to memory."

It was clear that Hall had tried to learn something about the area. "I read somewhere," he replied, "that several scholars have tried to study the Chippewa language and customs. I forget their names, except for a man named Henry Schoolcraft, after whom this county is now named." (Alger County, in which Munising and Grand Island are now located, was not split off from Schoolcraft County until 1885.)

Banta added that he believed that Schoolcraft had discovered the source of the Mississippi.

"I once met Mr. Schoolcraft," Clark stated.

"You did? How did that happen?"

"It is a long story. Perhaps I will tell you a little later."

"Anything you say," was Hall's amiable reply.

"Actually, you know, you are not quite right when you say that our people do not have written records," Clark continued. "It is true that we do not have a written language like you do, but we have our way of writing down important things. In this way we have written down several of our most important legends."

"Really? Where have you done this? Can I see these writings?" asked Hall.

"I would not tell you where they are even if I knew exactly, but I do not, since I have never been to this secret place. If the white men knew, one of them would damage or destroy the drawings, even though I can see that you wish to preserve such things. I will tell you only that the drawings are on stone cliffs and they are on the shore of this great lake. One of the artists was Shingwauk, or Little Pine, who lived on Grand Island before I was born. The drawings are still watched over by a group of our people called Sugwaundugahwini-newug, Men of the Thick Fir Woods."

"I hope that no one harms them, and that someday we will be able to study them," said Hall.

"Perhaps," said Clark.

Clark directed Hall to take the boat not to the nearest point of the island, the south end of the Thumb, but instead up the channel between the Thumb and the mainland, past Sand Point.

After a while they came upon the first outcropping of the cliffs, a rampart that was crumbling at the top. "Does that place have a name?" asked Banta.

"The white men sometimes call it the Solitary Buttress," said Clark. "It was struck by lightning last year and badly damaged. In our old Chippewa traditions it was considered a place inhabited by a spirit, and there are many stories about it, some of them going back centuries, long before the Chippewa were here. But for me it has a special significance because my mother fell to her death from it."

"That's terrible!" both the white men cried. "How did it happen?"

"It is a long story," said Clark, "and it influenced not only my life but the lives of all my boyhood friends. Do you really want to hear such a long tale, one that will perhaps be tiresome to you?"

The two visitors assured Clark that they would not find it tiresome, and urged him to tell the entire story.

Clark began, "My real name is Gashkiewisiwin-gijigong, Powers of the Air, but the white men found it awkward. When Mr. Williams, for whom I once worked, tried to write my Chippewa name in English, he said the result looked like the alphabet had gone over

PICTOGRAPH ON STONE

Niagara Falls and got torn up in the process. He started calling me Jim, and later it became Jim Clark. Everybody knows me by that name now, and has for over forty years. Even the other Chippewa here call me Jim.

"Actually," he continued, "Jim Clark is my third name. I was once Pangijishib, Little Duck, but my mother renamed me shortly before her death. The reason she renamed me is connected to the cause of her death."

"Please tell us about it, if you are willing to," entreated Banta.

"I would like to do so, but it is actually a song, and it is in Chippewa, so you would not understand it. Besides, my voice is not what it once was. It has been many years since I sang it to anyone, and it was best done with accompaniment by several of my friends, all of whom are now dead."

"Why don't you sing part of it in Chippewa so that we can get the idea of what it was like, and then tell it to us in English?" suggested Hall.

Clark fell silent, and the professor and the judge did not know whether he was likely to proceed or not. He closed his eyes and yet seemed to stare at the sky. After a few minutes he began slowly and silently to rock back and forth. Then, as the boat turned to the northwest toward Trout Bay, he started singing in a low voice, Chippewa words timed to a cadence and in a melody that Hall felt were from a ghostly museum. The two listeners were fixed in their places, even though they understood nothing, just as Governor Cass's men had been fixed in their places over sixty years earlier. As Clark warmed to his performance, his hands slipped over the gunwale of the boat. With one hand he began to beat time lightly against the side, and the other he slid back and forth, trying to make a sound like a woman's hand on the birchbark of a lodge. The beating hand slipped lower and lower on the hull of the boat until it was almost on the water line, picking up greater resonance from the proximity of the water. The beats did not sound quite like the old basswood log, but they reminded Clark of it. The song was part euphony, part ululation, and part vocable chart. Sometimes Clark stopped singing but continued to hum and caress the wood of the boat. Then he would begin again. He continued for about ten or fifteen minutes, and finally opened his eyes.

"Forgive me," he said, "I got a bit carried away and forgot that it all means nothing to you."

"It was ethereal," said Hall, "and deeply religious. I understand why you are a lay minister. You probably sing at your church."

"Not this song," replied Clark. "When we were first converted to Christianity we were told to sing Christian hymns, not pagan songs."

"There is a place for everything that is beautiful," observed Hall. "I teach my students about Greek and Roman literature even though it is pagan. Why don't you now tell us what your song was about?"

Clark began the tale, at first pausing frequently as if he were directly translating the song into English. He spoke of the old life of the Grand Island Chippewa, of their pleasure in the place and of the interruption effected by the criticism from their mainland tribesmen. He told of the war party, the Battle of the Cavern, and the hardships that followed. He described Autumn Duck, Sound of Wind in the Trees, Line of Thunder Clouds, Pearl-Whisper, and Little Red Flower.

The narration lasted a long time, and by the time it was finished the sailboat was nearing the beach in Trout Bay. The three men pulled the boat a few feet up the beach, then walked through the wild sweet peas and lichens and moss that come down to the sand, and then into the pine trees.

Walking to an open spot under the large pines, Powers of the Air said to the two visitors, "I recommend that you pitch your tent right here. It is where I was born, and if you stay here a while the place will come to mean a great deal to you, as it does to me. When you sleep here at night you will hear the most soothing of all sounds, the wind gliding through the pine needles above you. This sound cannot be described in words, not even Chippewa words. But it was loved by all the Chippewa who lived here, and my mother was named after it. And before too many nights, you will have an experience that is partly frightening and partly thrilling, when the wind shifts suddenly to the north and the smooth rustle of the pines turns into a rush and even a whistle. You will quickly reach for your blankets as the chill goes right through your tent. Then you will hear the crash of the waves on the beach and you will suddenly think of your boat, and dash out to save it from the waves. This place is alive, and as a living creature it will constantly present you with distinct and changing sights and sounds."

Clark said, with a little laugh, that occasionally when he came to Grand Island he had a few doubts about his adopted Christian faith. "Christianity tells me about the Holy Land," he said, "but that place is so very far away and seems foreign. I don't know whether they have pine trees there or not," he observed, "but they certainly don't have Lake Superior. And I doubt that they have anything like Trout Bay. When I lived here as a boy, my mother and father told me about our Chippewa religion—about the forest and water spirits; about the awesome and wonderful Kiwedin, the North Wind; about Manabozho, whom you white men turned into Hiawatha; and about all the lesser spirits of the cliffs and beaches. When I stand here and think about all that, it still seems more real than the virgin birth, the communal partaking of body and blood, the crucifixion and resurrection—all things that I preach on Sundays at our church. But don't worry; I know better than most people that the past cannot be brought back. I have known a simple past that will never be recaptured, and I do not favor even trying."

Clark helped Hall and Banta erect their tent and stayed with them that night. The professor set up his stove and oven and prepared biscuits. The judge found beans in his luggage for soup made with the water of Lake Superior. The three of them sat around the fire, eating soup and biscuits and talking of Grand Island and Clark's story. Banta made an unsuccessful attempt to call their guide Gashkiewisiwin-gijigong, but the old Chippewa man, after thanking him for the effort, said he preferred now to be called Jim Clark.

Hall wanted to know if Clark had ever read the poem *The Song of Hiawatha,* to which he had referred several times.

"Unfortunately, I never learned to read very well," said Clark. "But William Cameron, who used to be keeper of North Lighthouse, read it to me. It is a beautiful poem, and Longfellow put quite a few of our local legends into it. But from a Chippewa standpoint, it sounds very strange. The rhythm is nothing like our songs, as you will know if you remember what I sang to you."

"Longfellow, who died just last year, was a professor of European literature at Harvard College," said Hall, "and he was familiar with many poems in that literature. He borrowed the meter for his poem on Hiawatha from the epic poem of Finland, the *Kalevala*. It is called trochaic dimeter. No wonder the poem sounds strange to you. Some people even accused Longfellow of stealing the poem, what we call plagiarism. He has quite a few critics. Some say he was too sentimental."

"The song I sang to you was partly borrowed also," said Clark. "Nahbenayash, Line of Thunder Clouds, and I worked it out together, basing it on older Chippewa songs. I do not see anything wrong with that. And so far as sentiment goes, a life filled with both happiness and tragedy cannot be described properly without it."

"I agree with you," said Hall. "Our Baptist hymns are very sentimental, but I love to sing them. Many of them are borrowed from elsewhere, sometimes even from military songs."

Banta asked Clark if he had any children. "My wife, Pearl, died long ago," he said. "But we had several children, and now I have grandchildren and even one great-grandchild. One of my grandchildren studied at Albion College, in lower Michigan, and is an educated man. He is a real minister of the Methodist Church, not a lay minister like I am. His name is Thomas Thomas. Like his father, my son, his Chippewa name is Nahbenayash, Line of Thunder Clouds. We gave them that name in honor of the old Chippewa I knew in my youth whom I mentioned in my story and whose memory is so important to me."

They talked on into the evening. Clark invited Hall to come to the Indian Methodist church and give a guest sermon. Hall was interested in doing so but wondered if it would be appropriate. "Do you speak English or Chippewa at the church services?" he asked.

"The services are held in English," replied Clark. "All the Chippewa now know it. If we spoke in Chippewa many of the young people would not understand. For example, my own grandson, Thomas Thomas, who preaches at my church once a month, is losing his ability to speak Chippewa. But a few of the older members of the church, including me, still prefer to say some prayers in Chippewa, as we were originally taught. The Lord's Prayer, for example, is much more beautiful in Chippewa than in English. Whoever translated it into Chippewa—it was probably Father Baraga of L'Anse—did a wonderful job. The first line, 'Our Father who art in heaven,' is in Chippewa 'Our Father in heaven abiding,' *Nosinaun gezhigony abiyun*. It is too bad that you and your church members in Indiana do not know that prayer in Chippewa, because you would like it more."

"But Father Baraga was a Catholic," noted Hall. "You are a Methodist. Did he ever preach to you?"

Clark explained that in the early years, when the Jesuit priests had first come to the area, the Catholic faith had been the main influence. When the Protestants came they tried to run out the Catholics. The Chippewa did not see any difference between the faiths. So far as they were concerned the only religious difference that mattered was the one between their native religion and Christianity, and they were not sure they wanted Christianity. Some Chippewa, Clark continued, resisted the Christian religion for a long time. But when Christianity began to spread, sometimes it was Protestant, sometimes Catholic, and sometimes a mixture of the two. Father Baraga, a Catholic, and Reverend Pitezel, one of the early Methodist preachers, actively competed in the Munising and Grand Island areas. Eventually Pitezel prevailed, and most of the Chippewa there became Methodists, with Catholicism remaining strong farther west, closer to Baraga's home.

After talking a while longer the three men grew tired and decided to sleep. Clark showed Hall and Banta how to make soft beds by gathering some of the moss. Then they lay down in the tent, the old guide between the young professor and judge. All three lay there silently for quite a while but wide awake. The two white men thought they could guess what the Chippewa was thinking, and they were not far off.

Jim Clark had not slept on Trout Bay for many years. As he lay there and listened, the familiar sounds of the wind and the light surf were threads of memory leading to his youth, when he slept every night on the same spot. He was not Jim Clark anymore, not even Powers of the Air, but Little Duck. He recalled how, as a boy of twelve or thirteen, he watched his mother, Sound of Wind in the Trees, building a canoe between the lodge and the beach. First she had gone to gather birchbark and cedar, the main building materials. From the birches on the east side of Echo Lake she selected the best specimens, speaking to the trees:

> Lay aside your cloak, O birch-tree!
> Lay aside your white-skin wrapper.

From the stringy fiber made from bark of the white cedar she twisted *watab*, the cords to be used in sewing the canoe. Returning to Trout Bay, she set up a frame for the canoe, driving double posts in the ground to hold the bent pieces of wood in place. The arches in the center were larger and grew smaller toward the ends. The canoe was pointed at each end, but it was not symmetrical; instead it was slightly broader in the front than in the back, modeled on the shape of a fish. Each arched frame the length of the canoe was slightly different in shape, and Little Duck remembered that Sound of Wind in the Trees had a specific name for every frame and every other part of the canoe. To his dismay, he could no longer recall those names. The canoe had no keel, but the inside was covered with light slivers of cedar. The outside was sewn with a bone needle and cedar bast, then covered with resin.

The result was a one-person gossamer canoe that floated on the very surface of the water—

> Like a yellow leaf of autumn,
> Like a yellow water-lily.

The bottom of the canoe was so fragile that it could resist only bare or lightly moccasined feet. But one person could carry it easily on his shoulders from Trout Bay to Murray Bay and, with a few stops for rest, even to Echo Lake in the center of the island.

Somewhere in the midst of Clark's thoughts about his mother and the canoe, he fell asleep, as did the judge and the professor. When Hall awoke the next morning, as the first light penetrated the tent, he looked over and saw that Clark was gone. Pulling on his pants and shoes, he walked down to the beach to see if he could find the old Chippewa. About halfway toward the west end of the beach he saw Clark walking slowly along the shore. He had taken his shoes off and was splashing barefoot along the line where the water met the sand. Hall started off at a trot after him and soon began to catch up. But before he came close enough for Clark to be aware that he was being followed, he slowed to the old man's pace. Clark was walking very happily, somewhat stooped, but he had raised his face to the sky. When he was about two-thirds of the way toward the end of the beach, a point he seemed to be expecting, he lowered his face, looked to the northeast, smiled at something he saw, and then continued. When Hall reached the spot, he too looked northeast and saw, emerging from behind Trout Point (the northern point of the Thumb), a great arch in the cliff on the mainland, perhaps three hundred feet high.

This must be Grand Portal, thought Hall, about which he had read. When Pierre Radisson saw it in 1658 he wrote, "I gave it the name of St. Peter because that was my name and I was the first Christian to see it." This bid for immortality was not successful; later explorers knew it as Le Portail. When Henry Schoolcraft journeyed along these cliffs the day before he met Powers of the Air, he wrote in his journal, "It may be doubted whether, in the whole range of American scenery, there is to be found such an interesting assemblage of grand, picturesque, and pleasing objects."

Hall did not follow Clark farther but walked back along the beach to their camp, where the judge had rebuilt the fire and was frying bacon in the skillet. When Clark returned to the camp later, Hall said, "Your home is a very beautiful place, and I am only gradually learning about all of its attractions."

"Yes," said Clark, "but I am worried about its future. All along the Painted Rocks logging crews are now moving in. They are cutting the giants of the forests, the white pines that have grown here for centuries. On the other side of Grand Portal they are sliding the logs down the great sand dunes facing the lake. We Chippewa may have been pagans, but our religion honored nature."

Clark continued to say that many Chippewa believed the forest spirit was angry about what the whites were doing to forests and would find his revenge. He expressed a fear that the whites would eventually come to log the pines along Trout Bay, and asked if Hall thought they could be persuaded to leave them alone.

"I do not know," said Hall. "They have plenty of trees to log on the mainland now. It will be quite a while before they would think of expending the extra time and money to send logging crews to the island."

After breakfast Clark said that he would return to the mainland by himself so that the two campers could stay on Trout Bay.

"How will you do that?" asked Banta. "You have no boat, and it took almost two hours for us to get here over the water."

"One of my friends has fish nets set in Murray Bay, just a quarter of a mile walk from here," replied Clark. "He will be coming to

GRAND PORTAL

raise his nets later this morning. I will catch a ride back to town with him. If by chance he does not come, I will return here by noon and spend another evening with you."

"Would you like us to accompany you to Murray Bay?" Hall asked.

"No, thank you, I have used the short trail there all my life, and I will enjoy the walk," replied Clark. "But I hope to see you some Sunday evening at our church. It is in Indiantown, not far from my house. We would like for you to say a few words to us." He picked up his bag and jacket and left.

Hall and Banta set about completing their camp. They built a better fire circle, unpacked all their belongings, and readied their fishing gear and guns. About forty feet behind their tent they dug a latrine, placing a seat between two small pine trees. They were settling in for the summer.

In later years Columbus Hall camped in the same place. In 1888 he brought his daughter, Letitia Hall (later Carter), and she continued to come for eighty years. Her father returned annually until shortly before his death in 1926, more than forty years after his first trip. All of Professor Hall's eight children eventually camped there. Hall told Letitia ("Aunt Leta") the story of Powers of the Air, and she told her children, relatives, and friends, including me. Thus the Legend of Grand Island never died, although with time many of the details were forgotten or altered.

After the island was purchased by the Cleveland Cliffs Iron Company at the turn of the century, Hall moved his camp to the mainland, near Grand Portal, but his children later returned to Trout Bay. Eventually four cabins were built on Trout Bay, and today the summer resident of the one located where Powers of the Air suggested making camp is my own daughter, Meg Graham, a great-granddaughter of Columbus Hall. My wife, Patricia, first came to Trout Bay in 1937 at the age of two. The two small pines where Professor Hall placed his latrine are now large trees at the northwest corner of Meg's front porch.

That first summer Reverend Hall remembered his promise to visit Jim Clark at the Methodist church in Indiantown and speak to the congregation. But what would he say? He was afraid that his comments would seem strange or irrelevant to the Chippewa, a group to which he had never spoken before.

He kept thinking about Clark's view that living on Trout Bay sometimes made the ancient Chippewa religion more appropriate than Christianity. A deeply devout man, Hall found this bother-some. His concerns became greater the first time the wind shifted to the north, as Clark had predicted, and the great waves crashed on the beach. Standing there and looking at the scene, listening to the roar, Hall felt a subversive shiver. There *was* something pagan about the experience. He would think about this and perhaps say something about it when he went to the Indian church. Perhaps he could make a connection between the nature worship of the Chip-pewa and his Christian faith.

Before he went to college Columbus Hall had thought he might like to be a poet. He soon gave up the ambition when he found that his aesthetic aspirations were considerably greater than his abilities. He therefore decided to become a professor where he could teach, among other things, the poetry of Horace, Virgil, and Homer. But for his visit to the Indian church he resurrected his old aspirations.

On Sunday, July 15, 1883, Hall went to the Indian church, where Jim Clark welcomed him. Clark gave the regular sermon and spoke of the importance of the Christian faith for living a good life, es-chewing alcohol and thievery. After he had finished, he told the congregation that they were honored to have as a guest a distin-guished minister from Indiana, who would like to say a few words.

Hall began by telling the congregation how their minister, Jim Clark, had showed Judge Banta and him where to camp on Trout Bay. He then admitted that Reverend Clark had been correct when he told him that religious feelings could come not only from the written gospel but from nature itself. But this does not undermine the Christian faith, he continued; it actually fortifies it. (Hall real-ized that he said this with more conviction in the church than he had felt on the beach.) Right after he had witnessed his first north wind on Grand Island, he said, he had composed a short poem, which he now wished to read:

The North Wind

The north wind sweeps down the lake
And sets my nerves aquiver;
Exhausted energies awake
To thank the boisterous Giver.

He drives the fog upon the shore,
The leaden clouds come nearer,
Until with rumble and with roar
He sweeps the heavens clearer.

He paints the sky in deepest blue,
But paints the waters bluer,
Until your soul has caught the hue
Of something sweet and truer.

For all the soul within you leaps,
Emotions rise the higher,
And while the body lowly creeps,
In faith to God you're nigher.

Hall then sat down, hoping that he had managed to Christianize Kiwedin, the north wind of the Chippewa.

The congregation did not applaud, but several of the listeners politely nodded. Reverend Clark then said they would proceed to group meetings in which individuals could tell of their own religious experiences.

Later that night Hall described the meeting in his journal, a copy of which has been kept on Grand Island to this day: "After his sermon, by request, I spoke for 10 minutes and then we had a regular class meeting. There were three class leaders at the same time hearing the experiences of the various Indian men and women. When one would sit down, though two others were still speaking, they would start a song and those speaking seemed in no way disconcerted but went right on. It was to me a curious and interesting meeting."

The date of death of Jim Clark/Powers of the Air is unknown. It must have been between 1883 and 1890, probably before 1888. His name is not listed in the census rolls of the Chippewa of the Upper Peninsula taken at the end of the nineteenth century, although a number of his descendants are included in the 1908 rolls, residing in Munising, Bay Mills (near Sault Ste. Marie), and L'Anse. Two of his great-great-great-grandchildren, Marie Clark and Dolores LeVeque, live near Munising today. Both of them grew up in Indiantown (the name is still used), in separate houses a few hundred yards from the site of the Indian Methodist church, to which Columbus Hall came at the invitation of Jim Clark/Powers of the Air in 1883. Marie Clark lives today in the new Chippewa settlement near Wetmore, a few miles away. Dolores LeVeque (Minidemoya) still lives on the old home place in Indiantown, in a modern home

POWERS OF THE AIR

on a dirt road (Old Indiantown Road) a short distance from the ruins of the farmhouse where she was born.

Another of the descendants of Powers of the Air, Dwight (Bucko) Teeple (Wabun-Anung, or Morning Star), lives today in Sault Ste. Marie, where he is one of the leaders of the remarkable renaissance of Chippewa culture occurring there during the last twenty years. He was the model for the drawing on the previous page.

EPILOGUE: DECLINE
AND RENEWAL

THE FIRST TWO-THIRDS of the twentieth century were hard times both for the natural environment of the Upper Peninsula and for the Chippewa who were its original inhabitants. In the last decades of the century, however, there were signs of renewal.

By the first decades of this century almost all the virgin forests of the Upper Peninsula had been logged. As a result of lax game and fish protection, several species disappeared or almost disappeared: moose, wolverine, lynx, wolf, and lake sturgeon. In 1906 the Grand Portal at the Pictured Rocks crashed down into the lake with a roar so loud that people in Munising, fourteen miles away, heard it. The Chippewa remaining in the area said that the collapse of the great natural monument was a result of the anger of the forest spirit over the logging and desecration of the south shore of Lake Superior.

Grand Island itself was given a reprieve from this destruction by William G. Mather, president of the Cleveland Cliffs Iron Company, who purchased the island in 1900. Mather, a descendant of the New England divines Increase and Cotton Mather, was impressed by the beauty of the place and decided to create an island game preserve, a "Second Yellowstone Park," though in private, not public, hands. In pursuit of his conservationist goals he sought advice from two prominent friends: Stephen J. Mather (a distant cousin), the first director of the National Park Service; and Gifford Pinchot, of the U.S. Forest Service and eventually governor of Pennsylvania. William Mather went to great lengths to preserve the forests and wildlife of the island. He allowed no logging and only a few motor vehicles authorized by the company. Indeed, no one could set foot on the island unless invited by Mather. However, he often invited his friends to hunt, including Carter H. Harrison, the mayor of Chicago; and Oscar Mayer, the meat-packing mogul.

Mather was an avid preservationist, but he practiced a form of wildlife management that violated the accepted rules of today. He brought in many exotic species of animals and plants that were not adapted to the area—reindeer from Finland, elk from lower Michigan, brush hare from England, grouse from Sweden and Scotland, and American caribou and antelope. He also reintroduced moose. The importation project was a failure; many of the animals died out in the winter of 1902–03, although the albino deer and moose survived for some time. Simultaneously with his introduction of alien species, Mather tried to eliminate the native predators. At great expense he organized expeditions of several dozen men who hunted down and eliminated the wolves. One of the results was that the remaining deer and moose were not culled in a natural way, and disease, especially brain worm among the moose, was common.

Despite the flaws in his management of the island, so long as William Mather was alive the place was preserved in a relatively undeveloped state. He hired a gamekeeper to watch over the wildlife, planted Scotch pine on the cliffs where erosion threatened the coastline, and carefully preserved the old buildings constructed by Abraham Williams in the middle of the previous century.

A major exception to Mather's ban on construction on the island was a resort hotel and a few cottages, built between 1903 and 1909 on the south end of the island. In its early years the hotel was luxurious, and the board of directors of the Cleveland Cliffs Iron

Company sometimes held its meetings there. In later years it fell
out of favor with the privileged families of the Middle West, who
found other places with greater amenities. The hotel survived, usually on the edge of bankruptcy, for almost fifty years. Tourists were
permitted to come to the hotel and the cottages but not to bring
their vehicles onto the island. During the hotel's final years its management, desperately seeking to increase its attractiveness to tourists, decided to construct a bar, named the Gitchee Gumee, in the
basement of the hotel annex, which was the original home of Abraham Williams. By this time William Mather was dead, and his
strict rules against altering the appearance of the old buildings were
no longer enforced. In order to give the customers a better view of
Murray Bay, the hotel management bulldozed the small building
directly in front of the bar. That log structure was one of the original fur-trading cabins of the island. With it disappeared one of the
oldest buildings on Lake Superior. The morning after it was bulldozed, I visited the spot and pulled from the wreckage wrought-
iron fittings made by Williams (and perhaps by Powers of the Air),
as well as a beam with the initials of visitors to the trading post in
the early nineteenth century.

Mather always regarded Grand Island as a special place, and as
long as he was alive he gave it his idiosyncratic protection. After his
death in 1953 his company began to regard the island as just another
piece of its vast holdings in the Upper Peninsula. In the late 1950s
the public was allowed to hunt on the island for the first time, and
soon the deer, not accustomed to much hunting, were almost
wiped out. The eagles had already declined greatly in numbers as a
result of the nationwide use of DDT; their eggshells were so thin
that they broke before the young could hatch. The peregrine falcons, which had always lived on the north cliffs of the island, had
disappeared even earlier. The beaver on Echo Lake had never fully
recovered from the massive kill-off in the previous century, but a
few remained.

A few years after Mather's death the Cleveland Cliffs Iron Company, which ran large logging operations in the Upper Peninsula in
addition to its mining enterprises, announced that Grand Island's

virgin timber was to be selectively logged. The company purchased a tugboat and a lumber barge, which transported skidders and large trucks equipped with automatic loading machinery to the island. The trucks then returned to the mainland with white pine logs so enormous that sometimes only one log could be loaded on a truck. The company's timbermen estimated the age of the large trees at about five hundred years. They had difficulty finding a timber mill with a saw big enough to handle the logs.

In the early 1960s the company's foresters came to Trout Bay and marked the pines for logging. They drew diagonal blue stripes across the bark of the trees that were prime specimens. A few local people and summer residents of the island protested to the company's headquarters in Cleveland about the imminent harvest of the trees of Trout Bay. Somewhere in the top management of the company a person remembered Mather's love of Trout Bay and of the special characteristics of the place, and ordered that the trees of Trout Bay were not to be cut. In addition to Trout Bay a few other areas of virgin timber on the island were spared, including the old Indian and white burial grounds, and a few spots in the interior, along the coast, and near the northern end.

By the late 1950s the Lake Superior area had been well known and heavily exploited by whites for over a century. Surprisingly, however, not until 1958 did many know about the striking Agawa pictographs that Schoolcraft had mentioned in his writings and that Powers of the Air had described to Columbus Hall. Until that time the secret of the location of the sacred drawings, about ninety miles north of Sault Ste. Marie, had been kept by the Chippewa and a few white explorers. Selwyn Dewdney, a student of Native American art and religion, described in his diary his 1958 trip by boat along the rocky cliffs on the northeast shore of Lake Superior:

> We watched the Lake Superior shore go by: the long smooth
> curve of sand-edged Agawa Bay—calm in an offshore
> wind—the cluster of rocky islands off the promontory to
> the north behind which Agawa Rock lay hidden, and to the
> west the vast sweep of Superior. . . . At Agawa even in the

calm the water was restless beside the sloping ledge under
the sheer cliff. . . . I stared. A huge animal with crested back
and horned head. There was no mistaking him. And there, a
man on horse—and there four suns—and there canoes. I
felt the shivers coursing my back from nape to tail—the
Schoolcraft site! Inscription Rock! My fourteen months'
search was over.

Between 1958 and 1989, 117 pictographs were identified along
the shore, and more undoubtedly wait to be found.

The beginning of the renewal of the natural environment of the
Grand Island area came in 1966, when the federal government purchased more than forty miles of the south shore of Lake Superior
next to the island and created the Pictured Rocks National Lakeshore, a part of the national park system. The park area encompassed the Painted Rocks of the original Chippewa and the shoreline where Longfellow said that *The Song of Hiawatha* was based.
Fortunately, the area had never been settled except for a few vaca-
tion cabins, including one built by Columbus Hall and his family
after being forced off Grand Island when William Mather bought
it. By the time the national lakeshore was created second-growth
forest again covered the area. The coast is now well protected. The
person who travels today by boat from Munising to Grand Marais
sees the Pictured Rocks coast essentially as it has been for centuries,
with the exception of the collapsed Grand Portal.

In 1984 the Cleveland Cliffs Iron Company began seeking a
buyer for Grand Island, and its realtor initiated negotiations with
private developers, who spoke of converting the island into a vaca-
tion resort with hundreds of cottages. Alarmed at this prospect, my
wife and I helped organize the handful of families who had summered there for generations. The new Grand Island Association began actively seeking a buyer who would preserve the island in a natural state while keeping it open to hikers and visitors. Later that
year one of the islanders, William (Jeff) Geffine, contacted Tom
Offutt and Steve Morris of the Trust for Public Land about the possibility of purchasing the land and transferring it to the federal gov-

ernment. By incredible coincidence, Offutt's wife, Molly, was a great-great-niece of William G. Mather, and Tom and Molly already knew some of the history of the island. Offutt and Morris brought the possibility of preserving the island to the attention of Martin Rosen, president of the Trust for Public Land. In 1988 the Trust took out an option for the purchase of the island, hoping that federal legislation would enable the subsequent transfer. U.S. Congressman Dale Kildee, working closely with the U.S. Forest Service through his aide Larry Rosenthal, sponsored the legislation that permitted the purchase. In August 1990 the island was officially dedicated as a part of Hiawatha National Forest and thus saved from private development.

Since clear-cutting was never permitted on the island, it was still covered by forest, most of it second growth, in 1990. The federal legislation acquiring the island prohibited further logging and established the goal to "preserve and protect for present and future generations the outstanding resources and values of Grand Island." Local business people have pushed for the development of the island for tourists, including the construction of a new hotel, amphitheater, restaurant, boat marina, and bus transportation system. Conservationists, both local and national, have resisted this effort. The battle for the future of Grand Island continues.

Prohibition of the use of DDT helped in the restoration of eagles to the island, where several pairs were nesting by the late 1980s. In the early 1990s the Forest Service facilitated the return of peregrine falcons to the cliffs that gird most of the island. Moose were brought in from Canada to the Upper Peninsula and by the early 1990s had reestablished a permanent population with the few remaining native moose. Timber wolves, crossing over from Minnesota and northern Wisconsin, returned to the area at about the same time. Thus, although the ecological future of the Upper Peninsula and the area around Grand Island is not secure, the current trends promise a positive outcome so long as development can be controlled.

The Chippewa of the area experienced a decline far greater than that of the natural environment. Pushed by the whites either into

slums usually called "Indiantowns" on the edges of the towns or
cities or onto tiny reservations, most lived in conditions of appall-
ing squalor. A white man who in 1937 visited the Bay Mills Reservation near Sault Ste. Marie, where a number of the descendants of
the Chippewa around Grand Island had moved in the mid-
nineteenth century, observed that "to live in a poorhouse would
have been far better, for at least there they would have had something to eat."

In the Munising area the Chippewa chose at first to live along
the shore of the lake, near where the Washington School is now located, and later moved inland several miles, to the area called at first
Thomasville and, later, Indiantown. They chose this area both because it had land suitable for farming and because it was the site of
an age-old gathering spot of the Chippewa, a beautiful spring-fed
pond known as Tuhkib, Springdale. Led by Thomas Thomas,
grandson of Powers of the Air, they made a desperate attempt to
survive in difficult conditions. They earned their livings by farming, making baskets and moccasins, and selling maple syrup. The
main families in Indiantown in the late nineteenth and early twentieth centuries carried the names Ames, Bird, Blair, Blake, Brown,
Clark, Kishketog, Marshall, Sky, Thomas, and Walker. In many
cases their anglicized names had been assigned to them by the local
white storekeeper, W.A. Cox, in order to simplify his bookkeeping.

By the end of the nineteenth century the culture of these Chippewa was under heavy economic and political pressure from the
surrounding community. Several of their children, including Dan-
iel Sky, were sent by local authorities to the now notorious Indian
school in Carlisle, Pennsylvania, where they were required to wear
uniforms and were forbidden to speak their native language or ob-
serve native customs. They were expected to become whites. The
conversion was never complete, but several of them became excellent baseball players, and baseball became a favorite pastime in
Indiantown.

The Chippewa community near Munising continued for a
while the habit of holding periodic feasts, not Stone Pot Feasts, but
Hunting Feasts, and they transformed their rituals into Christian

ones. They were famous throughout the area for their storytelling and oratorical abilities, the passion of their hymn singing, and their generosity. Their melodic gatherings at Sand Point and Springdale sometimes attracted as many as five hundred people—both Indians and whites—from the surrounding areas. The last Hunting Feast was evidently held early in the twentieth century, but the hymn singing on Sand Point and at Springdale continued until much later.

Munising was too far north to become a prosperous farming area. Gradually many of the Indiantown community drifted away. Thomas Thomas eventually left for a Chippewa reservation in Wisconsin. Some descendants of Powers of the Air have remained to this day. Springdale Pond, located on their land, is still a Chippewa gathering place.

In the 1930s a road crew came to Sand Point, on the mainland just opposite Grand Island, to build a road to the life-saving station there. Sand Point was the place where the mainland Chippewa in the Pictured Rocks area usually buried their dead; in the mid-nineteenth century a number of former residents of the island had also been laid to rest there. The spirit-houses that were their final residences stood in sight of their older home across the channel. Many years later, in the 1980s, an elderly man who had been in the road crew told a member of the staff of the Pictured Rocks National Lakeshore that he wanted to make a confession about what had happened when he and his fellow workers built that road. They had, he said, run onto the old Chippewa cemetery. The spirit-houses were long gone, destroyed either by the elements or by vandals, but there were many skeletons just a few inches below the sand. The members of the road crew selected a few skulls to take home as souvenirs and threw the rest of the bones in a heap in a large ditch they dug not far from some small ponds. The old roadworker observed that he was now not far himself from the grave, and expressed the hope that his remains would not receive the same treatment that he had given to those of the Chippewa.

During Franklin Roosevelt's New Deal some legislation was passed that helped the Chippewa, but the real beginning of their revival came in the 1970s, when a new generation of Chippewa, or Anishinabeg, emerged. These men and women were often college educated, they took pride in their native heritage, and they learned how to benefit from several federal programs and laws that had been enacted by a white population feeling increasingly guilty about its treatment of Native Americans.

A leader in the renaissance of the Chippewa of the eastern Upper Peninsula was Joe Lumsden, who became chairman of the Sault Ste. Marie Tribe of Chippewa Indians (its official name) in 1976. His tribe included six historical bands: Drummond Island, Garden River, Grand Island, Point Iroquois, Sault Ste. Marie, and Sugar Island. Lumsden grew up in poverty in the Sault Ste. Marie area, speaking both Ojibway and English. An indifferent student, he managed to graduate from high school because of his interest in sports. He then joined the U.S. Marine Corps, where he first began to appreciate the importance of education. He realized that without it the Chippewa of his native region had no chance of emerging from the squalor that had plagued them in recent generations. At the same time he became deeply interested in the culture of the Chippewa. Upon his return home Lumsden attended Northern Michigan University in Marquette on the G.I. Bill, studying education, anthropology, and Native American history. He received his B.S. degree in 1967 and returned to his native Sault Ste. Marie with his wife and two children.

In the late 1960s and early 1970s the two areas near Sault Ste. Marie with the largest number of Chippewa were the Bay Mills Reservation and the Mar Shunk neighborhood. Both were poverty-stricken and afflicted by unemployment, alcoholism, and, increasingly, drug abuse. The Mar Shunk neighborhood was within the city limits of Sault Ste. Marie but had almost no urban amenities. The houses were tar-paper shacks, the streets were unpaved, there were no street lights, and the sanitation system consisted of outhouses or open sewers. Angered by these conditions, Joe Lumsden began an improvement campaign that continues today. He sought out allies among other young Chippewa and created an independent Native American power base that lobbied with state and federal authorities for assistance. One of the Chippewa who shared his vision was Robert Van Alstine, who helped establish an office of the Bureau of Indian Affairs in Sault Ste. Marie that was in step with the new movement. Van Alstine became education director of the office, a position he still holds. He began to work closely with a local college that is controlled by the Chippewa, Bay Mills Community College, and also with Lake Superior State University, which established a Native American Studies Department.

Bob Van Alstine and Joe Lumsden searched among the local Chippewa for other leaders in the growing movement. One of these was Dwight (Bucko) Teeple, a descendant both of Powers of the Air and of Mukwa-boam (Bear's Hump), brother of Omonomonee, the Grand Island chief who invited Abraham Williams to settle on Grand Island in 1840. Teeple's Chippewa name is Wabun-Anung, or Morning Star. His native name was given to him by his great-grandmother, Ellen Marshall (Maw-Dosh), related to the Marshalls and Clarks of the Grand Island/Munising area, who was midwife at his birth in 1948 in a cabin on the shore of Lake Superior. Like Lumsden, Teeple served in the military before returning to the Sault area and becoming involved in the growing Chippewa movement. Van Alstine encouraged Teeple to get a college degree, but Teeple believed he could not afford one. Van Alstine found a mobile home where Teeple could live cheaply while attending nearby Lake Superior State University, where he received a degree in social science in 1980.

Although Van Alstine and Teeple were allies in the movement, their styles were quite different. Van Alstine was a member of the Chippewa tribe, but he was also an administrator in the Bureau of Indian Affairs and was thoroughly modern in his method of operation. He specialized in getting education grants from the federal government and helping the tribe maneuver its way through the red tape of agencies and foundations. Teeple, on the other hand, became a traditionalist. He went on a "vision quest" on a small island in Lake Superior where he lived for four days without food or water. On his return to Sault Ste. Marie he participated in prayer services, naming rituals, sweat-lodge ceremonies, funerals, and native dances. He collected sacred plants and recorded cultural sites. In 1993 and 1994 he came to Grand Island and visited the spots where Powers of the Air was born and lived. Together Van Alstine and Teeple, the technocrat and the traditionalist, made a powerful combination, and they came to respect each other's strengths.

Many other Chippewa joined the movement, including Bucko Teeple's brother Allard, and by the 1980s the Sault Ste. Marie Tribe had become a powerful political force in the Upper Peninsula. On July 4, 1983, the Bay Mills Indian community opened the first tribally owned and operated gambling casino in the United States. Since that time the economic activity of the Chippewa communities has mushroomed. Today, the Sault Ste. Marie Tribe of Chippewa Indians has a multimillion-dollar annual budget and is the largest employer in the Upper Peninsula of Michigan, larger than the Mead Paper Company, the Cleveland Cliffs Iron Company, or the Kimberly-Clark Paper Company, all of which have large operations there. The tribe's current chairman is Bernard Bouschor of Sault Ste. Marie. Its officers include George Nolan, a relative of Sophie Cameron; Barbara Pine, a descendant of Shingwauk, the warrior who left Grand Island in protest against its pacific policies; and Victor Matson, Jr., who represents the Grand Island and Munising areas. Matson is the grandson of Charles Matson, a Chippewa fisherman who at the turn of the century took Columbus Hall by boat to his camps on Grand Island and Beaver Beach near the Pictured Rocks.

The Sault Ste. Marie Tribe currently operates more than thirty businesses, including a construction company, a telephone company, health centers, a community school, restaurants, a large hotel, and several motels. It has completely reconstructed the Mar Shunk area of Sault Ste. Marie, paving the streets, building a model village to replace the tar-paper shacks, and erecting nearby a large hotel, casino, and restaurant. It has formed an Economic Development Council, made up of both whites and Native Americans, and currently chaired by a white man. Unemployment in the area has decreased fifty percent in ten years. As far-flung members have re-enrolled, the tribe has grown to 23,000 and is one of the largest in the nation.

The new prosperity has allowed the tribe to expand its economic activity throughout the eastern Upper Peninsula. Near Munising, not far from the Indiantown where Columbus Hall found Powers of the Air in 1883, a new Chippewa settlement has been constructed. The neat and attractive houses are placed along streets with Ojibway names. The hovels from which the Chippewa moved were burned to the ground. One resident of these new homes, at 123 Na Me Guss (Trout) Street, is Marie Clark, great-great-great-granddaughter of Powers of the Air. Dolores LeVeque, another great-great-great-granddaughter, lives about three miles away. Several miles east of the new settlement stands a tribal center for the Munising/Grand Island area, which provides basic health services. The old Indiantowns near other cities in the Upper Peninsula are being similarly renovated.

Despite the enormous progress made by the Chippewa of the Upper Peninsula in recent years, many problems remain. Anyone who visits the older Chippewa communities, such as the Bay Mills Reservation, can still find poverty, alcoholism, and unemployment. Furthermore, the prospect of a continuing cash flow from the casinos at its current rate is uncertain, since it seems unlikely the Native Americans' near-monopoly on gambling will last. The Upper Peninsula Chippewa, aware that their current economic advantage may be temporary, are trying to strengthen their educational facilities to build for the future, diversifying their economic activities, and continuing to strengthen a new and sturdy pride in their culture.

I finished a draft of this manuscript in early September 1994, working on the dining table in North Light, where Jim Clark/Powers of the Air first heard Longfellow's poetry read to him by William Cameron. I packed up the manuscript and took it down to the beach, then launched my boat, steering first to the north and then around the northeast tip of the island, named Point Gallant after Joseph Gallant, a consummate woodsman of French-Canadian descent who helped clear the site for the lighthouse. During the ten-mile trip down the east side of the island I saw an eagle and a peregrine falcon flying over the cliffs. To the east I watched the afternoon sun reflecting off the Pictured/Painted Rocks. The south shore of Lake Superior stretched as far as I could see, but not a single building or manmade structure was visible, although I spotted a large white tourist boat from the Pictured Rocks Cruise Lines making its way along the cliffs. I admired the promontory still called Grand Portal even though the arch is no longer there. The cliff that remains still looms high above the lake, and the bare outline of the arch can still be seen. I recalled the legend of the Wawabezowin, the magic swing on the Painted Rocks cliffs. I took my boat across the entrance of Trout Bay and stared at the beach where Little Duck made his morning runs. Rounding Trout Point, I headed toward Munising, passing the Solitary Buttress where Sound of Wind in the Trees fell to her death. Instead of heading directly in to the Munising City Dock, where I could have walked three blocks to the post office, I headed west toward the much smaller dock at Powell's Point, where my car was parked.

As I crossed Munising Bay I could see, straight ahead on the southern end of the island, Abraham Williams' house and the small beach in front of it where Powers of the Air sang to Lewis Cass and Henry Schoolcraft. To the left, in the town of Munising, behind the trees I could just make out the Superior Health Haven Nursing Home, where Julia Cameron had died just a few weeks earlier. In her last year—her ninetieth—she reverted to her childhood. One evening the staff of the nursing home discovered that she was missing and notified the town policeman, who found her walking toward downtown Munising. When he asked her where she was going she replied that she was going to Grand Island to pick blueberries with her Chippewa mother, Rosalie Larmond.

After landing at Powell's Point I drove to the post office in Munising and mailed the manuscript. Then I turned my car westward again, but instead of returning to my boat I decided first to take another look at Powers of the Air's face on the rock near Au Train, thirty-five years after Harry Powell had first told me about it. I drove through the hamlet of Christmas, where the Chippewa tribe had just opened another small casino, and continued another five or six miles to the east end of the beach at Au Train, where I pulled my car off the road into the small roadside park now located there. In the parking lot stands a metal plaque commemorating the sighting of Lake Superior by Etienne Brulé in 1622. Nowhere in the pocket-sized park operated by the State of Michigan is there a sign indicating that a petroglyph is nearby.

I walked down to the beach and along its edge toward the low rock wall several hundred feet away. I remembered exactly where to go, and soon I stood on the same spot as I had in 1959. I looked up and was relieved to see that Powers of the Air remains on the cliff face. No sign or marker is here either, and several other people on the beach were oblivious to the face that had attracted me there, even when I stood staring at it. However, not far from the face a Romeo has carved into the rock a heart, inscribed "Lisa 'n Guy." Tourists have carved their initials into the cliff at several other spots.

Despite the graffiti, the place is still strikingly lovely. No buildings are anywhere near it, although a short distance up the little brook, the state authorities have built a small wooden bridge so that people can get from the parking lot to the beach more easily. I walked inland up the brook and soon reached the waterfall where Cass's guides bathed, but to get there I had to cross M-28, a two-lane Michigan highway that covers the old Chippewa trail.

Returning to the face in the cliff, I saw that it has deteriorated since I first saw it. The chin has fallen away, but it is unclear whether a vandal or the weathering of nature was responsible. The face is low enough that in the worst storms it is washed by the waves of the lake. I imagine that during the winter the face is often covered by ice. It seems extraordinary that it has survived at all. The top part of the face is still in good shape, and I could see the two eagle feathers.

Below the face, where the shoulders would have been, lies a horizontal white stripe, very weathered, with another short vertical stripe below it. At first I thought the stripe was old lichens, but then I saw that someone had once, many years earlier, painted an arrow indicating the face so that people would notice. Thankfully, the arrow has weathered so quickly that it is as muted as the face itself. The face continues to be protected only by its anonymity. Few people know about it even yet.

I thought about the manuscript and realized that when this book is published, people will learn about the carving in the rock and will probably come to this spot looking for it. Will they damage it? Will I contribute to the destruction of the story that I have tried to rescue? As I returned to my boat and headed back to Grand Island, these disturbing thoughts ran through my mind. I could only hope that by telling the legend of Powers of the Air I have conveyed something of the strength, courage, and integrity of the Grand Island Chippewa, and that my readers may come, as I have, to view Powers of the Air's fading visage as an icon to his culture. Perhaps by increasing awareness of and appreciation for this rock carving, I will have helped in a small way to preserve that culture.

SOURCES

PRIMARY SOURCES

Interviews and conversations on Grand Island from 1954 to 1969 with Letitia Hall Carter, daughter of Columbus Horatio Hall, provided some of the important elements of this story. Columbus Hall's diaries (vols. 1–6) and letters, together with the diary of his son Nelson Clarence Hall (vols. 1–3), located in the North Light Collection, Grand Island, supplied other details. Some of Columbus Hall's diaries are now on deposit in the Burton Historical Collection, Detroit Public Library. The poem that Hall read to the congregation in Powers of the Air's church was later published in the *Baptist Outlook*, Indianapolis, August 19, 1897. Conversations in the 1960s and 1970s with John Lezotte, employee of the Cleveland Cliffs Iron Company from 1914 to 1975 and supervisor of the island for several decades, were helpful, as were talks with his brother Tony Lezotte, mayor of the town of Munising and tugboat captain. Members of old Grand Island families with whom I had valuable discussions included Harry Powell, Lewis DesJardins, and Aaron Powell. Interviews from 1959 through 1985 with Julia Cameron, native Chippewa speaker and granddaughter of William Cameron, keeper of North Light, Grand Island, enlightened me further. William Cameron's logs, complete with some of the poetry quoted in this book, can be found in Record Group 26, National Archives, Washington, D.C. A letter from Julia Cameron to John O. Viking, December

6, 1943, describing her family, is on file in the Marquette Historical Society, Marquette, Michigan.

It was a thrill to meet in the summer of 1994 two great-great-great-granddaughters of Powers of the Air/Jim Clark, Marie Clark and Dolores LeVeque. The existence of Marie Clark was brought to my attention by Vivian Price of the Munising–Grand Island Tribal Center. Marie Clark's grandmother Martha Blair was from another old Grand Island Chippewa family, the descendants of Antoine Blair. At the time I found Marie Clark, the manuscript for this book was almost finished. Upon meeting her, before I could say anything about the Legend of Grand Island, she launched into a description of her ancestor Nahbenayash (Line of Thunder Clouds), whom Powers of the Air had described so proudly to Columbus Hall in 1883. When Marie Clark learned that I was writing a book about her ancestors, she telephoned her cousin Dolores LeVeque, who soon supplied more information. Several generations of their family carried the name Nahbenayash, including Jim Clark's son, known as Chief Nahbenayash, or Quay-quay Cub; and Thomas C. Thomas, Jim Clark's grandson, the Methodist minister and graduate of Albion College. On the basis of conversations with Dolores LeVeque and Marie Clark, it appears likely that the original Nahbenayash—the Chippewa who was too old to participate in the war party but who helped Powers of the Air to write the song about the battle—was a close relative of Autumn Duck, possibly his uncle or even his father. This relationship would help explain why the name Nahbenayash then passed through several generations of descendants of Jim Clark/Powers of the Air.

On August 15, 1971, a taped interview with Belle Dorothy Carr Kroupa, a great-great-granddaughter of Powers of the Air, was made by her granddaughter Mari Beth LeVeque. The tape is now in the possession of Dolores LeVeque. In the tape Mrs. Kroupa described the early history of Thomasville, or Indiantown, in considerable detail. People of Chippewa descent sometimes still gather at Springdale. In July 1994 a name-giving ceremony was performed there for Darwin Kroupa, brother of Dolores LeVeque, who received the name Wa-Wa Geshi (Deer).

Conversations and correspondence in 1994 with Keith Cameron, Bay Mills Indian Reservation, Brimley, Michigan, helped establish links between the Grand Island/Au Train Cameron family and the Camerons currently living near Sault Ste. Marie.

Conversations in 1993 and 1994 with Bucko Teeple (Morning Star), of the Office of Planning and Research, Sault Ste. Marie Tribe of Chippewa

Indians, Sault Ste. Marie, provided insight into the fate of the Chippewa community since the days when Grand Island was one of the centers of their activity. Robert Van Alstine, a Chippewa tribal member who is also Educational Director in the Bureau of Indian Affairs, Sault Ste. Marie, helped me to decipher the Durant Roll and other nineteenth-century census data on Upper Peninsula Chippewa; he also told me much about the history of the Sault Ste. Marie Tribe of Chippewa, of which the Grand Island branch were (and are) members. The Durant Census Rolls of 1870 and 1910 are on deposit at the Bayliss Public Library, Sault Ste. Marie, along with relevant papers from the Superintendency of Indian Affairs, 1841–51, and the Office of Indian Affairs, 1824–81.

The 1673 map of Lake Superior by Father Jacques Marquette, showing "Les Grandes Isles," is in the Jesuit Archives, Maison Monserrat, St. Jerome, Quebec, under the current care of Father Robert Toupin.

A vast collection of Henry Schoolcraft's papers has been microfilmed and is on deposit in the Library of Congress, Washington, D.C. See *Henry Rowe Schoolcraft: A Register of His Papers in the Library of Congress* (Washington, D.C.: Library of Congress, 1973). Items that related particularly to this story include Container 73, Reel 58 ("Mishosha, or the Magician of the Lakes," and "The Charmed Arrow"); and Container 74, Reel 58 ("Wawa be zo win, or the Swing"). Schoolcraft notes here that "Grand Island in Lake Superior was the traditional home of Mishosha as well as of Kaubina." The record of Schoolcraft's visit to Grand Island in 1820, when he heard Powers of the Air sing the story of the lost Grand Island warriors, is available in several different editions, including Mentor L. Williams, ed., *Narrative Journal of Travels through the Northwestern Regions of the United States Extending from Detroit through the Great Chain of American Lakes to the Sources of the Mississippi River in the Year 1820, by Henry R. Schoolcraft* (East Lansing: Michigan State College Press, 1953), especially pp. 104–10. Schoolcraft described another visit, on July 21, 1822, in his *Personal Memoirs of a Residence of Thirty Years with the Indian Tribes on the American Frontiers: With Brief Notices of Passing Events, Facts, and Opinions, A.D. 1812 to A.D. 1842* (Philadelphia: Lippincott, Grambo, 1851), p. 102. Other Schoolcraft papers are dispersed around the country, but relatively few mention Grand Island; the best listing is in Richard G. Bremer, *Indian Agent and Wilderness Scholar: The Life of Henry Rowe Schoolcraft* (Mount Pleasant, Mich.: Clarke Historical Library, Central Michigan University, 1987), p. 434.

There are several different accounts of the Battle of the Cavern, both in written and in barely surviving oral traditions. The best known is the one given by Schoolcraft in his *Narrative Journal,* pp. 109–10. Other accounts were given by James Duane Doty, Lewis Cass, Charles C. Trowbridge, and David B. Douglass, all of which derive from the song of Powers of the Air on June 21, 1820. See James D. Doty, "Tale of the Thirteen Chippewas," *Detroit Gazette,* January 12, 1821, excerpts of which are published in Schoolcraft's *Narrative Journal,* Appendix F; also a letter from Lewis Cass to John C. Calhoun, February 2, 1821, reprinted in part in ibid., Appendix C; Charles C. Trowbridge's journal of the expedition of 1820, partially reprinted in ibid., Appendix G; and Sydney W. Jackman and John F. Freeman, eds., *American Voyageur: The Journal of David Bates Douglass* (Marquette: Northern Michigan University Press, 1969), p. 50. There is much disagreement over the details of the battle, and some indication that Schoolcraft misunderstood Powers of the Air on important points. Some accounts describe the Grand Island Chippewa as going to meet the Sioux alone and in secret, while others have them joining with the mainland Chippewa, as in my version. Some accounts even maintain that Grand Islanders were not involved in the battle at all, but only Chippewa from the Leech Lake area of Minnesota. A source in which the Grand Islanders are omitted, and the besieged warriors described as being Pillagers from the Leech Lake band (the mainland Chippewa who, in my version, fled the scene before the battle), is William W. Warren, *History of the Ojibway People* (St. Paul: Minnesota Historical Press, 1984), p. 388. According to local Grand Island lore, this account is incorrect and was contrived by the Pillagers to cover up their own cowardice in fleeing the scene. It is not at all clear which, if any, of these various accounts is correct. For the purpose of the story in this book, I have accepted the Grand Island version.

The policy of the federal government toward Indians in the first half of the nineteenth century, including the promotion of their removal west of the Mississippi River, is documented in many primary sources. One that pays particular attention to the Chippewa of the Upper Peninsula is "Report of the Commissioner of Indian Affairs," Office of Indian Affairs, Department of War, Washington, D.C., November 16, 1842; and "Sub-Agency for Chippewas," La Pointe, Wisconsin, September 30, 1842; both in *Message from the President of the United States to the Two Houses of Congress at the Commencement of the Third Session of the Twenty-seventh Congress,* December 7, 1842 (Washington, D.C.: Gales and Seaton, 1842), especially pp. 370 ff. and 404 ff.

The competition between the Catholic priest Frederic Baraga and the

Methodist minister John Pitezel for the souls of the Grand Island Chippewa, whom both visited in 1845 and 1846, is described in Frederic Baraga, letter to Archbishop Milde, Archbishop of Vienna, January 24, 1846, Bishop Baraga Archives, Marquette; and John Pitezel, *Lights and Shades of Missionary Life* (Cincinnati: Western Book Concern, 1861), pp. 112–16.

The visit of Henry Wadsworth Longfellow's friend Louis Agassiz to Lake Superior before Longfellow began writing *The Song of Hiawatha* is described in Agassiz, *Lake Superior: Its Physical Character, Vegetation, and Animals, Compared with Those of Other and Similar Regions* (Boston: Gould, Kendall and Lincoln, 1850).

On February 26, April 12, and April 14, 1849, Longfellow met with the Chippewa chief Kahgegagahbowh, or George Copway. At that time Copway gave Longfellow a copy of his autobiography, which included a description of the south shore of Lake Superior and a visit to Grand Island. Longfellow noted in his journal that Copway "described very graphically the wild eagles teaching their young to fly from a nest overhanging a precipice on the Pictured Rocks of Lake Superior." See Samuel Longfellow, ed., *Life of Henry Wadsworth Longfellow: with Extracts from His Journal and Correspondence* (Boston: Ticknor, 1886); also George Copway (Kahgegagahbowh), *The Life, History and Travels of Kahgegagahbowh* (Albany, N.Y.: Weed and Parsons, 1837).

An autobiography that contains much information about the Chippewa of the Lake Superior region is *A Narrative of the Captivity and Adventures of John Tanner (U.S. Interpreter at the Saut de Ste. Marie) during Thirty Years Residence among the Indians in the Interior of North America*, ed. Edwin James (Minneapolis: Ross & Haines, 1956). A valuable early map of Grand Island is S. Seibert, "Map of Munising on Grand Island Bay in Lake Superior in Schoolcraft County, State of Michigan, 1850–55," J.M. Longyear Library, Marquette, Michigan. A book of photographs that was helpful to Abigail Rorer in her drawings of Chippewa artifacts is *The Art of the Great Lakes Indians* (Flint: Flint Institute of Arts, 1973). A study showing that the Trout Bay area was inhabited by human beings well over two thousand years ago is Elizabeth D. Benchley, Derrick J. Marcucci, Cheong-Yip Yuen, and Kristin L. Griffin, "Final Report of Archaeological Investigations and Data Recovery at the Trout Point 1 Site, Alger County, Michigan," University of Wisconsin–Milwaukee Archaeological Research Laboratory, Report of Investigations, no. 89, February 1988.

The warfare between the Chippewa and the Sioux is attested in many primary sources. One of the more interesting is the series of drawings

made by the famous Sioux leader Sitting Bull (c. 1832–80) while a prisoner
of the U.S. Army at Fort Randall, South Dakota (1881–82), depicting his
greatest victories as a warrior. The drawings, which came from the private
collection of R. Anderson, are on display at the Plains Indian Museum of
the Buffalo Bill Historical Center, Cody, Wyoming. One of these depicts
a battle with fifty Chippewa in which Sitting Bull, as a young brave in a
war party of some two hundred Sioux, killed a Chippewa antagonist. This
battle must have taken place around 1850, and as such is evidence that the
Chippewa-Sioux warfare continued to a late date. In fact, according to
some sources, it continued to the time of the American Civil War.

The old log Indian church in Indiantown, where Jim Clark invited Columbus Hall to come to preach in 1883, is still in existence, though barely
recognizable. In 1915 it was moved from Indiantown to East Munising,
covered with clapboards, and became a private residence visible today
from the main highway, near the Anna River.

The Ojibway dictionary used for this book is Frederick Baraga, *A Dictionary of the Ojibway Language* (1878; reprint, St. Paul: Minnesota Historical Society Press, 1992), a reprint with new material of Baraga's original
dictionary of 1878. Some reference was also made to Richard A. Rhodes,
Eastern Ojibwa-Chippewa-Ottawa Dictionary (New York and Berlin:
Mouton de Gruyter, 1993); Coy Eklund, *Chippewa (Ojibwe) Language
Book* (Petoskey, Mich.: Indian Hills Trading Post, 1991); and George
Monroe Campbell, *Campbell's Original Indian Dictionary of the Ojibway
or Chippewa Language* (Minneapolis: Campbell Publishing, 1940).

SECONDARY SOURCES

The only other book about Grand Island is Beatrice Hansom Castle, *The
Grand Island Story*, ed. James Carter (Marquette: Marquette County Historical Society, 1974). Though in many ways a remarkable and helpful
source, it does not cover the story of Powers of the Air, describing only his
meeting with Lewis Cass and Henry Schoolcraft in 1820. Another valuable though incomplete secondary source on Grand Island is Norene
Roberts, "Cultural Resources Overview and National Register Evaluation
of Historic Structures, Grand Island National Recreation Area, Michigan:
Final Report," U.S. Department of Agriculture, Hiawatha National Forest, 1991.

A slender but useful book is Olive M. Anderson, *By the Shining Big-
Sea-Water: The Story of Pictured Rocks 10,000 B.C. to 1966 A.D.* (Munising:
Bayshore Press, 1989). In this source, based in part on interviews and un-

published sources, Anderson describes the meeting between Columbus Hall and Jim Clark in 1883; she refers to the latter as "Indian Jim."

An interesting and helpful description of the Chippewas in the Munising area is Faye Swanberg, "In the Very Beginning," in *Who Were Those People?* ed. Charles A. Symon (Munising: Alger County Historical Society, 1982), pp. 87–99.

In the vast literature on the Native Americans of the upper Midwest, the fur trade of the seventeenth, eighteenth, and nineteenth centuries, and the history of relations between the Native Americans and the French, British, and American governments, I found a number of sources that were particularly helpful. Charles E. Cleland, *Rites of Conquest: The History and Culture of Michigan's Native Americans* (Ann Arbor: University of Michigan Press, 1992), contains a large bibliography. A book that relates much Chippewa history, primarily of the members of the tribe who lived to the west of the Grand Island band, is Edmund Jefferson Danziger, *The Chippewas of Lake Superior* (Norman: University of Oklahoma Press, 1978). Older, even classic sources existing in recent reprint editions include William W. Warren, *History of the Ojibway People* (St. Paul: Minnesota Historical Society Press, 1984); and Johann Georg Kohl, *Kitchi-Gami: Life among the Lake Superior Ojibway* (St. Paul: Minnesota Historical Society Press, 1985). A good discussion of Chippewa lifeways can be found in James A. Clifton, George L. Cornell, and James M. McClurken, *People of the Three Fires: The Ottawa, Potawatomi, and Ojibway of Michigan* (Grand Rapids: Michigan Indian Press, Grand Rapids Inter-Tribal Council, 1986). A detailed chronology of events involving Native Americans in the Upper Peninsula (including a reference to the Battle of the Cavern and several other events mentioned here) is Russell M. Magnaghi, *A Guide to the Indians of Michigan's Upper Peninsula* (Marquette: Belle Fontaine Press, 1984). A source with a rich bibliography and information on the Chippewa is Bruce G. Trigger, *Northeast: Handbook of North American Indians,* vol. 15 (Washington, D.C.: Smithsonian Institution, 1978). Information about construction of Chippewa lodges and burial rites came from Bucko Teeple and also Frances Densmore, *Chippewa Customs* (St. Paul: Minnesota Historical Society Press, 1979). Much information on the Grand Island/Munising/Au Train area came from Charles A. Symon, ed., *Alger County: A Centennial History, 1885–1985* (Munising: Bayshore Press, 1986). Also helpful was Thomas Overholt and J. Baird Callicott, *Clothed-in-Fur and Other Tales: Introduction to Ojibway World View* (Washington, D.C.: University Press of America, 1982). An entertaining, immense, and not altogether reliable history of the central Upper Peninsula of Michigan is C. Fred Rydholm, *Superior Heartland: A Backwoods History,* 2 vols. (Marquette: privately published, 1989).

A source showing that Lewis Cass and Henry Schoolcraft used whiskey to prevail with the Upper Peninsula Chippewa, despite their frequent denunciations of this practice by others, is Bernard C. Peters, "Hypocrisy on the Great Lake Frontier: The Use of Whiskey by the Michigan Department of Indian Affairs," *Michigan Historical Review,* 18, no. 2 (Fall 1992), 1–13. Peters has also written other articles giving the original Ojibway place-names for Lake Superior locations; see, for example, "The Origin and Meaning of Place Names along Pictured Rocks National Lakeshore," *Michigan Academician,* 14 (Summer 1981), 41–55; and "The Origin and Meaning of Place Names along Michigan's Lake Superior Shoreline between Sault Ste. Marie and Grand Marais," *Michigan Academician,* 26 (1994), 1–17, which furnishes in n. 1 the titles of Peters' earlier works on the topic.

An intense and somewhat overstated effort to cement the relationship between Longfellow and Schoolcraft in the writing of *The Song of Hiawatha* is Chase S. Osborn and Stellanova Osborn, *Schoolcraft-Longfellow-Hiawatha* (Lancaster, Pa.: Jaques Cattell Press, 1942). An account of the Johnston family of Sault Ste. Marie, into which Schoolcraft married, is Deidre Stevens Tomaszewski, "The Johnstons of Sault Ste. Marie: An Informal History of the Northwest, as Portrayed through the Experiences of One Pioneer Family," Bayliss Public Library, Sault Ste. Marie, n.d.

For more information on Abraham Williams, see Sarah Smith, "The Life of Abraham Williams (with special emphasis on Grand Island)," J.M. Longyear Library, Marquette, 1978.

White fur traders often complained that warfare among Native Americans interrupted the fur trade, but in fact they promoted that warfare by constantly pushing the native trappers to expand their territory to obtain more furs. See Harold Hickerson, *The Chippewa and Their Neighbors: A Study in Ethnohistory* (New York: Holt, Rinehart and Winston, 1970).

A source describing a grave near the Tahquamenon River in 1835 as that of Mukwa-boam (Bear's Hump), "the brother of the present chief of the Grand Island Indians," who was at that time described as Omonomonee, is Larry B. Massie, "Guardian of the Great Lakes' Graveyard: A History of Whitefish Point and Its Lighthouse," Michigan Council for the Humanities, Sault Ste. Marie Tribal Headquarters, Sault Ste. Marie, July 1991. Anna Maria Williams, daughter of Abraham Williams, recalled that

Omonomonee was in Sault Ste. Marie when her father arrived there in 1840 and invited Williams to settle on Grand Island (see Castle, *The Grand Island Story*, p. 32). Another source says that in 1849–50 the Grand Island chief Omonomonee (also spelled Monomonee) "had recently moved from Grand Island" to Whitefish Bay. The report continues: "many of the Grand Island band moved to Naomikong in the 1840's." Naomikong is between Whitefish Point and Bay Mills (Helen Hornbeck Tanner, "Report: United States of America vs. the State of Michigan, No. M26-73 C.A., U.S.D.C.," Western District of Michigan, Northern Division; located in headquarters, Sault Ste. Marie Tribe of Chippewa Indians, Sault Ste. Marie).

A good description of the construction of the canal at Sault Ste. Marie is Irene D. Neu, "The Building of the Sault Canal: 1852–1855," *Mississippi Valley Historical Review*, 40, no. 1 (June 1953), 25–46.

The apostate Grand Islander Shingwauk (Shinguacose), Little Pine, was a fascinating character who deserves a separate study. His life and exploits are still remembered among his descendants around Sault Ste. Marie. See, for example, the recollections of Dan and Fred Pine in Thor Conway and Julie Conway, *Spirits on Stone: The Agawa Pictographs* (San Luis Obispo, Calif., and Echo Bay, Ont.: Heritage Discoveries, 1990), pp. 69–74. For a mention of Shingwauk's burning his collection of birchbark pictographs and other artifacts before his death, see Kohl, *Kitchi-Gami*, p. 384.

A fine source on Chippewa pictographs, including some information on Shingwauk, is Selwyn Dewdney, *The Sacred Scrolls of the Southern Ojibway* (Toronto: University of Toronto Press, 1975). Also see Selwyn Dewdney and Kenneth E. Kidd, *Indian Rock Paintings of the Great Lakes* (Toronto: University of Toronto Press, 1967). Helpful on Lake Superior is Margaret Beattie Bogue and Virginia A. Palmer, *Around the Shores of Lake Superior: A Guide to Historic Sites* (Madison: University of Wisconsin Press, 1979).

An informative source on the recent activities of the Chippewa in the Upper Peninsula is their monthly newspaper, *Win Awenen Nisitotung*, Sault Ste. Marie.

A discussion of the geology of the Grand Island/Pictured Rocks area, with photographs of clastic dikes on Grand Island, is W. Kenneth Hamblin, *The Cambrian Sandstones of Northern Michigan*, Michigan Department of Conservation, Geological Survey Division, Publication 51 (Lansing, 1958). Also helpful is John A. Dorr, Jr., and Donald F. Eschman,

Geology of Michigan (Ann Arbor: University of Michigan Press, 1970). A description of the exceptional length of the beaver dam on Echo Lake, Grand Island, is G. Evelyn Hutchinson, *A Treatise on Limnology,* vol. 1 (New York: John Wiley and Sons, 1957), p. 146. A source pointing out that the sand spit connecting the Thumb to the rest of Grand Island may be the longest tombolo on the Great Lakes is Kim Alan Chapman, Michigan Natural Features Inventory, Lansing, Michigan, letter to Loren Graham, September 24, 1984, North Light Archive, Grand Island. A shorter but wider tombolo is at Pequaming, near L'Anse.

For a discussion of William G. Mather's effort to preserve the island, see Laurence Rakestraw, Fred Stormer, and Christopher R. Eden, "William G. Mather and the Grand Island Game Preserve," *Journal of Forest History,* July 1977, pp. 156–63; "Grand Island Forest and Game Preserve," J.M. Longyear Collection, Marquette County Historical Society, Marquette; and H.J. Tompkins, "A Forest Working Plan for Grand Island," Manuscript, Yale Forestry Library, New Haven, 1905.

A discussion of historic structures on Grand Island is MacDonald and Mack Architects, Ltd., "Historic Structure Report," U.S. Forest Service, Minneapolis, 1993. See also "Description of Buildings, Premises, Equipment, Etc. at Grand Island Light-Station," Light-House Establishment, Department of Commerce and Labor, June 28, 1909, North Light Archive, Grand Island.

On the acquisition of Grand Island in 1990 by the federal government, see H.R. 1472, 101st Cong. 1st sess., March 16, 1989, sponsored by Representative Dale Kildee of Flint, Michigan.

On the history of Munising, see Rev. Emil J. Beyer, "History of Munising," John M. Longyear Library, Marquette, 1983. A description of how the Grand Island Chippewa used flaming torches in Murray Bay to catch fish at night is in L. Masson, *Les bourgeois de la compagnie du Nord-Ouest* (Quebec: A. Coté, 1889–90).

ACKNOWLEDGMENTS

The research for this book was largely a process of listening to other people, either through conversation, if they were alive, or through reading. I am deeply indebted to these preservers of the fragments of the story. Most of them have already been named in the text.

The one person who more than any other is responsible for the book is my wife, Patricia Albjerg Graham. Without her I would never have come to Grand Island, and without her family I would not have learned the story. She has been a witness to events, a critical reader of manuscript drafts, and a delightful and invaluable companion for forty years.

Our daughter, Meg, has proved to be not only a superb lighthouse restorer but also an invaluable literary critic; this book bears her stamp as well.

Although I started gathering the material that eventually went into this book thirty-five years ago, I did not actually decide to write it until 1992, when a friend of our family, a fourteen-year-old German boy named Marci Weingart, asked me to tape-record some stories about Grand Island. The first version of my account of Powers of the Air was born on that tape. A few months later, my editor, Howard Boyer, who had earlier worked with me on several books on the history of Russian science, yet wanted me to write a book on Lake Superior and Native Americans, persuaded me to lend it to him. After listening to it he came to Grand Island with his wife, María Eugenia Quintana, learned the story on his own, and convinced me that I should write the book. Howard is the sort of editor whom every author dreams of finding but rarely does. He commits himself totally to his authors, reads the drafts, makes suggestions, and becomes a personal friend. Working with him is a privilege.

It was Howard who suggested that Abigail Rorer do the illustrations for the book. Abigail came to Grand Island and visited the spots where the events in the book take place. As a result of her talent and dedication, the artwork possesses the immediacy and authenticity that come only through personal familiarity.

Laurie Burnham at Shearwater Books has taken great interest in the project and made valuable suggestions that have improved the book. Ann Hawthorne, who edited the text with extraordinary care, deserves special thanks.

While gathering material for the book I have incurred a large number of debts. Members of the Chippewa tribe have given me hours, even days, of their time. Julia Cameron, Marie Clark, Bucko Teeple, and Bob Van Alstine, whose roles I have already mentioned, were essential to the project. Bucko's brother Allard also came to Grand Island and offered his assistance. Dolores and Art LeVeque took me on a tour of historic sites near Indiantown. Dolores also read the entire manuscript and helped in many other ways. Victor Matson and Vivian Price of the Munising/Grand Island area Chippewa center read the manuscript and offered useful comments. Sharon Fleming, a Native American student-worker at the library of Bay Mills Community College and the Chippewa Cultural Center, generously gave us her time when Abigail Rorer and I visited the Bay Mills Reservation, answering our questions and showing us a collection of artifacts that provided both background and ideas for illustrations. Alan Kamuda, staff photographer, *Detroit Free Press*, provided me with photographs of Chippewa dress.

I have visited the Alger County Historical Society in Munising many times over the years and am grateful to its long-term president, Isabella Sullivan, for assistance. Faye Swanberg, an accomplished local historian of the Munising area, saved me from a number of factual errors. Linda K. Panian and Rosemary Michelin of the Marquette County Historical Society assisted me in finding materials in the ample collections of their library and archive. Alice Paquette of the Houghton County Museum helped me to learn about the ancient portage route through the Keweenaw Peninsula. Regis Walling, editor-archivist of the Bishop Baraga Archives in Marquette, Michigan, gave me the benefit of her unparalleled knowledge of the Baraga papers, unerringly directing me to Baraga's references to Grand Island. Barbara Moore of the Superior Health Haven Nursing Home told me of the events in the last months of Julia Cameron's life. Father Robert Toupin, Jesuit Archives, St. Jerome, Quebec, answered my questions about Father Marquette's early maps of Lake Superior. Grant Petersen, Superintendent, Pictured Rocks National Lakeshore, read the manuscript and offered many helpful comments. Teresa Chase, Ranger, Munising District, Hiawatha National Forest, introduced me to Bucko and Allard Teeple and deserves additional thanks for being a good steward of Grand Island. Two other members of the Hiawatha National Forest staff who have been particularly helpful are Dick Anderson and Wally Jurinen.

During the years in which information for this book was collected, many people in the cities of Munising and Marquette gave assistance. Chief among them are Ted Belfry, Peter Benzing, Sheila Bonner, John Carr, Chuck Coolman, Bob Cromell, Bruce Cromell, Al Gruetzmacher, Tom Hall, Bart Nikkari, Bob Oas, and the Van Landschoot family of Van Landschoot Fisheries.

The families who summer on Grand Island—the Barnetts, Beichs, Carlsons, Ericksons, Fredericksons, Geffines, Grahams, Jossis, Kanes, Leutheusers, Needhams, Poppers, Predls, and Thomases—are in the habit of helping one another at all times. I thank them for dozens of small acts of assistance, as well as for good friendship. Special thanks go to Hobby Beich, Bill Thomas, and Kathie Carlson for their constant care.

Many academic colleagues have offered suggestions, and several read part or all of the manuscript. Leo Marx and Merritt Roe Smith, my colleagues in the Program on Science, Technology and Society at the Massachusetts Institute of Technology, rendered both stylistic and substantive help. At an early stage Leo encouraged me to go ahead with the project; Roe provided additional sources and helped with integrating the story into American history. Russell Magnaghi, professor of history at Northern Michigan University, gave me the benefit of his knowledge of Native American studies. Irene Neu, my former colleague at Indiana University, helped me find information on the building of the Sault Ste. Marie Canal. Anton Struchkov of the Institute of the History of Science and Technology in Moscow advised me on environmental issues.

Bonnie Burg engaged me in discussions about the focus and title of the book, helping me to make essential improvements. My brother Eldon and his wife, Vicky, have come up with a myriad of ingenious suggestions about how to live in an isolated spot such as Grand Island. Jerilyn Edmondson, my assistant at MIT, was a part of the project from beginning to end, helping in numerous ways. None of these people is responsible, of course, for the flaws that may remain.

INDEX